Further Praise for *Beyond Science Standards*

It is easy to think of modern natural science as an experimental, laboratory-based and sometimes algorithmic undertaking. *Beyond Science Standards* tackles that perception head-on, prompting readers to engage in the rich diversity of scientific thinking and methodology across contexts and disciplines. Charles "Kip" Ault demonstrates convincingly how scientists use their imagination to engineer systematic methods of inquiry that respond to the questions they ask.

—Marianne Achiam, Associate Professor of Science Education,
University of Copenhagen, Denmark

As I read *Beyond Science Standards*, I found myself smiling, remembering my days as a middle school science teacher, trying to navigate the dual complexities of science and teaching, while trying to foster a classroom in which kids felt safe to explore, tinker, and feel joy at asking and answering questions. I realized that my enjoyment of the book came from, the stories of actual classrooms and real kids and their teachers' genuine efforts to do good work. The stories nicely illustrate the joy, messiness, and wonderful reality of science classrooms. As a reader, you feel transported into these spaces, eagerly awaiting updates about the students and the science. Rarely are people, including researchers, given such a whole picture of daily classroom life. Rather than the atomized examinations of a single feature of classroom life, this book takes a panoramic perspective, inviting readers to experience the sights, sounds, and smells of classrooms. This book illustrates how classrooms can become special as the teachers and students enjoy working and learning together. If nothing else, know that I am now drifting off into nostalgic memories of John J. Pershing middle school in Houston, where I taught students who are now doctors, musicians, dancers, lawyers, and realtors. What we share—our classroom experiences—are nicely captured in this book.

—David Stroupe, Ph.D., Associate Director for STEM Teacher Education in the CREATE for STEM Institute and Associate Professor of Teacher Education, Michigan State University

"Pesky Pillbugs" is a great chapter in *Beyond Science Standards*—I feasted on it before starting my work emails this morning. I love the juxtaposition of high-level thinking/rigor with whimsy. Kip Ault's notion of "guiding imagery" is the main workhorse of our Embryology unit—we found that such images capture the imaginations of our bilingual, emerging scientists.

—Sunny Sue Chang Jonas, Ed.D., Assistant Principal, Chicago Public Schools

Beyond Science Standards

Beyond Science Standards

Play, Art, Coherence, Community

Charles R. Ault Jr.

ROWMAN & LITTLEFIELD
Lanham • Boulder • New York • London

Book epigraph: "Arguments don't change minds, stories do." Excerpt from A *Pilgrimage to Eternity: From Canterbury to Rome in Search of a Faith* by Timothy Egan, copyright ©2019 by Timothy Egan, page 273. Used by permission of Viking books, an imprint of Penguin Publishing Group, a division of Penguin Random House LLC. All rights reserved.

"Worms" section in chapter seven, "Broken Sticks," excerpted from "Irritating Worms" in *Do Elephants Have Knees? And Other Stories of Darwinian Origins* by Charles R. Ault, Jr. Used by permission of the publisher, Comstock Associates, an imprint of the Cornell University Press, © 2016.

"Sketchy Animals" section in chapter twelve, "Vivid Canyons," abridged from "Out on a Limb: Sketching at the Zoo to Compare Limb Anatomy," in *Do Elephants Have Knees? And Other Stories of Darwinian Origins* by Charles R. Ault, Jr. Used by permission of the publisher, Comstock Associates, an imprint of Cornell University Press, © 2016.

Published by Rowman & Littlefield
An imprint of The Rowman & Littlefield Publishing Group, Inc.
4501 Forbes Boulevard, Suite 200, Lanham, Maryland 20706
www.rowman.com

6 Tinworth Street, London SE11 5AL, United Kingdom

British Library Cataloguing in Publication Information Available

Library of Congress Cataloging-in-Publication Data

Names: Ault, Charles R., Jr., 1950- author.
Title: Beyond science standards : play, art, coherence, community / Charles R. Ault Jr..
Description: Lanham : Rowman & Littlefield, [2021] | Includes bibliographical references and index.
Identifiers: LCCN 2021017262 (print) | LCCN 2021017263 (ebook) | ISBN 9781475859966 (cloth) | ISBN 9781475859973 (paperback) | ISBN 9781475859980 (epub)
Subjects: LCSH: Science—Study and teaching—Social aspects—United States. | Science—Study and teaching—Standards—United States. | Experiential learning—United States. | Interdisciplinary approach in education—United States. | Teaching—Philosophy.
Classification: LCC Q183.3.A1 A84 2021 (print) | LCC Q183.3.A1 (ebook) | DDC 507.1/073—dc23
LC record available at https://lccn.loc.gov/2021017262
LC ebook record available at https://lccn.loc.gov/2021017263

For India Wood,
partner in paleo, world class adventurer, and my best student ever
and
Neal Maine,
consummate community activist, unmatched
teacher, and inspiring ecologist.

Arguments don’t change minds; stories do.

—Timothy Egan

Contents

Foreword by Jack Hassard xiii

Acknowledgments xvii

Introduction: The Cheshire Cat's Grin 1

PART I: PARADIGM LOST 9

1 Descartes' Dream 11

2 Complicated Truths 21

3 Discipline Diversity 29

4 Menacing Mudflats 37

PART II: PLAYFUL EXPLORATION 49

5 Wavy Elephants 51

6 Binary Banjos 65

7 Harmful Haste 75

8 Serious Whimsy 85

9 Salted Alcohol 93

10 Squirrel Tales 99

PART III: AESTHETIC EXPRESSION 107

11 Skull Sockets 109

12 Crowned Molars 125

13 Hell's Pig 137

14 Vivid Canyons 147

15 Fossil Imagery 155

16 Inspiring Invertebrates 163

17 Clawed Wings 177

18 Pesky Pill Bugs 187

19 Flashy Plumage 195

PART IV: CONCEPTUAL COHERENCE 203

20 Poetic Rocks 205

21 Vast Moments 217

22 Storied Geology 225

PART V: COMMUNITY PURPOSE 235

23 Simple Automata 237

24 Egg Balloons 247

25 Competing Forecasts 257

26 Wicked Extinctions 265

27 Harvesting Oysters 277

28 Caring Communities 285

Conclusion: Paradigm's Progress 293

Index 305

About the Author 323

Foreword by Jack Hassard

When I was invited to write the foreword to the book you hold in your hands, I felt honored. After I started looking at the manuscript, I realized I was reading a bellwether book in the field of science teaching. I've never read such a fascinating, beautifully written work. Ever. This remarkable manuscript, *Beyond Science Standards*, uncovers new ground upon which we can change the way we teach and learn science.

Like many of you, the digital world of email and social media is the mode of communication that brings us together in productive ways without actually meeting in person. That's how Charles "Kip" Ault and I came to know each other and exchanged science education research and writing we've done over the past thirty years. Because we both taught courses in geology and the earth sciences to a range of audiences, our career paths have crossed in several ways.

Kip was professor of science education at Lewis & Clark College for twenty-four years. There he developed and coordinated the science teacher education program. During the same decades, I was professor of science education at Georgia State University where I directed the Global Thinking Project, and the Technology, Engineering, Environment, Math and Science (TEEMS) AmeriCorps program for secondary science teachers. Kip introduced my book and website, *Minds on Science*,[1] to his Master of Arts in Teaching candidates in the early 1990s. Later, I read several of his science education research articles and his book *Challenging Science Standards* to explore notions of science teaching for my weblog. We are now emeritus professors of science education at our respective universities.

Teaching and learning are under great stress not only because of an older and rigid system of control but because of radical effects of Covid-19 on how and where students learn. In my view, the act of writing *Beyond Science Standards* has taken what Rollo May[2] calls "creative courage." May goes on to say that creative courage is the discovering of new forms, new symbols, new patterns upon which a new society can be built. Courage is embedded in Kip's act of questioning the present, old paradigm of science teaching which is based on the authoritarian spray of standardization that has spread harm and inflicted damage on America's schools over the two decades. The Next Generation Science Standards is part of this paradigm.

Beyond Science Standards charts a creative path to a new paradigm of science teaching and learning. It shares an amazing journey of stories and experiences in classrooms that will be familiar to you. Descriptions of people, events, and situations are hallmarks of qualitative methods. Kip has filled his book with playful, aesthetic, meaningful, and compelling stories about learning in which the value of context and the needs of students reign. The book is a treasure chest of examples from a new paradigm. Kip's book is also fun to read, with whimsically named stories such as Wavy Elephants, Binary Banjos, Skull Sockets, Crowned Molars, Hell's Pig, Vivid Canyons, Flashy Plumage, and Wicked Extinctions.

These stories, and others, take us into real classrooms to experience science content selected and designed with the needs and aspirations of students in mind. As Kip explains, the sciences taught in schools should "demonstrate their value in the personal, social, and political lives of young students." In one story, students in Ms. B.'s science class explore visual and aesthetic expressions of paleontology and geology to learn the evolutionary descent of modern horses and how fossils of horses are witness to climate change, as well as changes in landforms.

Of particular interest here is Kip's point that paleontology cannot exist without illustration. Illustration, drawing, and painting are each art forms that can be the heart of learning science. Kip stresses that art forms in science are not enhancements or enrichments. "They are a means to teach science beyond the standards." You will be thrilled to read the chapters on aesthetic expression in Kip's conception of science teaching.

Indeed, I would suggest that his conception of science teaching is drawn from the science of Leonardo da Vinci. Kip creates a new paradigm rooted in Leonardo's mind. *Beyond Science Standards* describes a holistic and ecological worldview that, according to Fritjof Capra,[3] is not unlike Leonardo's. Leonardo developed a solid body of science, but his science could not be understood without his art, nor his art without his science.

Walter Isaacson[4] and Fritjof Capra wrote separate biographies of Leonardo. In their biographies, they explain that Leonardo's scientific explorations informed his art. Capra says that for Leonardo "painting is both an art and a science—a science of natural forms, of qualities, quite different from the mechanistic science that would emerge two hundred years later." For Kip Ault, paleontology cannot exist without illustration, and he shows how art can be the center of methodology—and not only in paleontology. Art can be at the center of learning science.

Believing that science educators have perpetuated the process skills approach, and a generic approach to scientific inquiry, he argues that the nature of inquiry for a paleontologist, for example, is different than a chemist. In particular, he sees that using a generic approach without creating a disciplinary context for inquiry shortchanges the students and removes a sense of true inquiry.

In the second decade of the twenty-first century, we are enduring the worst pandemic in a century. Millions of youths organized a worldwide protest over the climate crisis in 2020. Climate change has fueled fires not only in the U.S. Western states but in Australia and other countries. Death Valley had its hottest day on record in July 2020 at 54.4°C (130°F). Some cities are becoming uninhabitable. Hurricanes have increased in intensity and have wreaked havoc on the Gulf States, especially due to extreme flooding and tsunami-like wave action.

Ever since Rachel Carson[5] and Frances Oldham Kelsey asserted their scientific findings, warning of the use of DDT as a pesticide and thalidomide as a drug, respectively, there has been a well-funded science denial community that pushes back against scientific findings that are truths that some groups do not want to accept. During the past four years, the previous administration served as the poster child for science denialism by creating havoc with the health care system, failing to respond adequately to Covid-19, while continuing to roll back environment regulations that were enacted to protect the air, water, and land.

Believing that the purpose of education is to prepare citizens for lives of social responsibility in a democratically governed polity, Kip reaches out to the science education community to claim that our present practices of teaching and routines of selecting what to teach will not help our students achieve that end. He concludes that immersing students in "scientific diversity" can be a journey uncovering aspects of ourselves and the universe promising immense pleasure and joy. Kip Ault has written the book that I've been waiting for.

Jack Hassard
Professor emeritus, Georgia State University
Author of *The Art of Teaching Science*
Weblog (artofteachingscience.org)

NOTES

1. Jack Hassard, *Minds on Science: Middle and Secondary Methods* (New York: HarperCollins, 1992).
2. Rollo May, *The Courage to Create* (New York: W. W. Norton & Company, Inc, 1975), 21.
3. Fritjof Capra, *The Science of Leonardo* (New York: Knopf Doubleday Publishing Group, 2007).
4. Walter Isaacson, *Leonardo Da Vinci* (New York: Simon & Schuster, 2017).
5. Rachel Carson, *Silent Spring* (New York: Houghton Mifflin Company, 1962).

Acknowledgments

In the summer of 2019, science educators Dawna Bell and Cathie Alder invited me to join them on a panel to discuss the future of science teacher education at the annual conference of the Oregon Science Teachers Association. I wish to thank them profusely for drawing me back into the fray. My notes from that event became the first outline of *Beyond Science Standards*.

Wu Jinshan and Chen Lijuan of Beijing Normal University called upon me to brush off course syllabi, step out of retirement, and teach in China. I thank Joseph Novak, one of my Cornell mentors, for linking me to them and am grateful for the honor of being part of Jinshan's "Teach Less/Learn More" team.

For a close reading and careful edit of the manuscript, I wish to thank Glenellen Pace, my language arts and drama education counterpart at Lewis & Clark College for many years. Her touch improved my writing immensely.

Once again, I turned to Kristen Hall-Geisler of Indigo Editing to count on professional feedback from an independent source. Kristen provided the editorial help leading to publication of my previous books, *Challenging Science Standards* and *Do Elephants Have Knees?* Her "reader response" insights prompted revisions to *Beyond Science Standards* that I, being so close to the manuscript, could not have made on my own.

John Settlage's encouragement of my writing has mattered to me greatly. Participation in the Science at the Crossroads conferences inspired by his leadership has been crucial to honing my arguments and gaining confidence to put them in writing.

Tom Koerner and Carlie Wall, my editors at Rowman & Littlefield, along with Arun Rajakumar at Deanta, have stayed with me throughout the process, most of which unfolded during the months of pandemic. Tom was my ever-responsive, anchor of normalcy, as I embedded myself in research and writing.

I have kept the names of elementary and secondary classroom teachers anonymous. The stories of their classrooms constitute the centerpiece of the book. To "Ms. B.," an exemplary secondary science teacher, my deepest thanks for her help with the story about teaching the evolution of the horse. To the fantastic "Mr. Link" and his two extraordinary fourth-grade team colleagues who went well beyond the call of standards to engineer binary banjos, I am forever indebted. They invited me into their classrooms for several weeks, making me feel I was back in 1975 teaching at The Children's School in Colorado Springs. I have never witnessed a team more open to innovation and dedicated to creativity. And finally, to "Mr. D." and everyone at Arbor School of Arts and Sciences, thank-you for showing me how project work is done to perfection.

For several weeks, the specimens from the John Day National Monument Horse Fossil Study Kit took up residence in the guest room of our home. The monument's Park Ranger, and talented watercolor artist, Michelle Ordway, guided me through the kit's organization and resources. Her contributions polished the paleontological sections of the book.

My son, Toby Rollin Ault, walked me through the pilot phase and first results of the promising Bootleg Balloon project in rural New York. I had planned to observe middle schoolers in action in 2020. Of course, the pandemic postponed their activities. The story, nevertheless, remains emblematic of teaching ideals for the twenty-first century.

No one has read and improved more versions of the ever-changing chapters than my beloved wife, Phyllis Campbell Ault. For many years she was the coordinator of the Northwest Tribal Educators Alliance through the Affiliated Tribes of Northwest Indians. I have tried to incorporate a small part of her work in the book as an example of teaching science that matters to a community.

I am humbled by Jack Hassard's foreword and express sincere gratitude for his contribution to the book. A collection of voices inspired *Beyond Science Standards*. My thanks to them all.

Introduction

The Cheshire Cat's Grin

Rare, indeed, is the high school science teacher who begins a course by pronouncing "Disunity dominates science!" More likely, students hear the teacher say something to the effect, "More important than remembering the facts is to appreciate 'the' scientific method." In an opening gambit of instruction, no one expects a science teacher to dwell on the incommensurability of scientific ideals from one field (or historical period) of inquiry to another.

Deemed central to the progress of civilization, grasping the "general character" and "habits of mind" of the sciences—historically and popularly encapsulated in the phrase "the scientific method"—reigns as the preeminent value of teaching science in K-12 schools. Assuming unity masks the sciences' vital differences. Novel ideas extend the frontiers of science in many directions. They flow from imagination's confrontation with seemingly inexplicable phenomena and intractable problems. "Disunity" is neither a pejorative term nor a perspective that discounts the integration of disciplines essential to tackling novel challenges with new technologies.

School science's historic devotion to unity—science-as-process and other generic flavors of inquiry—back-burners the intrinsic joy of knowing the sciences in their full glory. Embracing diversity as a hallmark of science focuses attention on the ways inquiry adapts to the nature of phenomena. Context reigns. Experience accumulates. Understanding satisfies. Solutions have value. From this standpoint, inculcating students in the habits of mind common to all the sciences is a recipe for obscuring nature's astonishing kaleidoscope of phenomena.

Schooling advances by immersing students in respect for scientific diversity. More must follow: teaching the worth of a variety of subjects chosen for their potential to contribute to social progress. Social and historical circumstances constantly frame new problems. Timely and local topics merit stature, displacing universal, timeless, standardized ones presumed equally important for all students to learn everywhere.

ADVANCING SCIENCES

An unexamined assumption of universal method, embedded in the adjective "scientific," reflects a paradigm under assault since the mid-twentieth century. The conditions scientists valued in the conduct of their research rose in social and political prominence in the wake of the extraordinary contributions from the sciences to the prosecution of the Second World War. The practices of free and flourishing inquiry, transparent and trusted, cast doubt on the ubiquity and validity of a singular notion of scientific method. Rather than one method of problem-solving to rule them all, intellectuals cast "the" scientific method as myth.

Scholars at Harvard University suggested case studies in the history of science as the best means to separate the wheat from the chaff, subject by subject, and thus dispel the myth. In the context of this work, a graduate student and course instructor, Thomas Kuhn, hatched telling insights into the nature of conceptual change within a subject. He referred to fundamental epochs of change as "scientific revolutions" that transformed not only methods of investigation but also the criteria for warranting new claims. Change on this scale meant that a "paradigm shift" had advanced the discipline. In such epochs, out with the old, in with the new created conflicting and irreconcilable ways of thinking—so said the history of science, according to Kuhn.

Just as a plural society benefits from the distinct and different contributions from the cultural perspectives of its many members, so, too, does science profit from the diversity of enterprises assembled to solve perplexing puzzles and chart promising futures. The sciences depart from each other in order to advance through appropriately disciplined efforts that stretch the boundaries of experience and deepen the languages of explanation, matching them with methods of investigation. *Disunited*, the sciences develop their particular expertise; *reassembled*, they converge to take on the grand challenges of the present epoch.

PLACING TRUST

From a disciplinary perspective, twenty-first-century science is quite ecumenical; the data of interest to ecologists and climatologists can be found in surprising venues, as suggested by this message from a new graduate student in paleoclimatology:

> Great news! I've just been accepted into graduate school in geology with the opportunity to work on a terrific project. The professor would like someone to do a photographic survey of the Lower Colorado River documented 150 years ago in journals and watercolor paintings. The aim is to compare habitats and channels of today with those from the past within the context of reconstructing climate trends in western North America.[1]

Climate scientists test hypotheses on virtual earths—while society runs analogous experiments (the effects of carbon emissions) on the real one. A good climate model computes probable scenarios of future climate from a simulated earth. Populations of virtual earths, constructed from observations, physics, and computations, exist within models. The modeled earth is known through observations sampled on various timescales, from ancient past to present, obtained from specific places, but not all. Slartibartfast, the creator of planets in Douglas Adams's fanciful *The Hitchhiker's Guide to the Galaxy*, would appreciate the task.

No vast population of earth exists in reality, for example, from which to select a sample at random for an experimental treatment, such as manipulating carbon dioxide concentrations in the atmosphere. Climate science must practice other ways to warrant claims and earn trust, commensurable with the nature of the case: one earth, complexly modeled.

Models adhere to basic principles of physics and chemistry, and successful reconstruction of past climates from historical data validates modeling. The resolution scale of real-world details sets limits to climate modeling, of course. As sampling resolution and completeness improve, the constraints on uncertainty narrow. Modeling generates credible, disturbing, scenarios.

Accepting the need to improve modeling does not invalidate their reasonableness. Climate science is rational and convincing, even if it does not adhere to a stereotyped scientific method.

Misplaced criteria for judging different sciences imperil our future because misunderstanding of how a particular science gets done empowers political interest groups determined to spread doubt about the trustworthiness of discomforting claims.[2] Claiming that computer modeling fails to meet the criteria of verifiable experimentation undermines public confidence, for example. The judgment that a claim meets an appropriate ideal—valued in one context, dismissed as irrelevant in another—is key to placing trust in a science. No single paradigm effectively subsumes all disciplines.

ADVANCING SCHOOLING

In the wake of the stature of Harvard's paradigm-shifting understanding of the sciences themselves, instructional aims in science education pivoted, crystalizing to become teaching nature of scientific thinking within major disciplines. Science educators began to acknowledge that scientific enterprises functioned in splendid and distinct ways.

In the latter decades of the twentieth century and into the twenty-first, educators doubled down on making continuous inquiry the hallmark of school science, yet often failed to heed philosopher David Hawkins's warning, "The mind equipped with method and no content is no more than the grin of a Cheshire Cat, an absurdity of misplaced abstractions."[3] Making inquiry skills or scientific processes a content domain in its own right, a clear case of "misplaced abstractions," did little more than reinvent the myth of the scientific method.

An inconvenient truth about the sciences—"Different disciplines have different starting points and different goals. Subject matters may be conceived in vastly different ways"—continues to nag on the coattails of educational reform.[4] The three-dimensional synthesis of science in the Next Generation Science Standards—scientific practices, core ideas, and crosscutting concepts[5]—shifts the rhetoric but not the paradigm of school science. Even an interpretation of the standards honoring disciplinary diversity with integrity and depth has shortcomings. The Cheshire Cat's grin remains.

Valuing Context

If not the scientific method, the nature of major disciplines, or the three dimensions of the current standards, what should be the top-level organizer for teaching *Beyond Science Standards*? The "Paradigm Lost" (part I) to science no longer governs the paradigm of science education. Sciences selected for school need to demonstrate their value in the personal, social, and political lives of young citizens. The achievement of social value, neither the myth of singular science nor the nature of inquiry discipline by major discipline isolated from social and historical context, should occupy the pinnacle of instructional planning.

When the achievement of social value sets the context, subjects gain classroom footholds by virtue of demonstrating their worth. Teaching the basis for trusting the contributions of a science becomes imperative. No single science sets the standard for all of the sciences. Not unless and until science teaching fully rejects the myth of singular science can students learn to appreciate scientific rationality in its multiple guises.

Beyond standards lie realms of untapped curriculum diversity, knowledge selected for its worth as a public good, education responsive to society's needs, and inquiries evaluated according to appropriate criteria. In diversity lies strength, the strength to solve societal problems, build community capacity, and boost citizen participation.

Appreciating Stories

Pathways leading beyond standards pass through the gates of creativity and context. Playful exploration and aesthetic expression add to the adventure. Conceptual coherence and community purpose guide the journey.

Stories of "Playful Exploration" (part II) honor Anna Botsford Comstock's joyful approach to the study of nature and draw inspiration from David Hawkins' conception of "messing about." A young child reflects sunlight off a submerged mirror and observes a "rainbow" cast on the wall. Two middle school boys record "bathroom noises" and use a computer program to make them sound like bird chirps and whale songs. By making two-string banjos to send messages in binary code, fourth graders confront the challenge to engineer a device demonstrating how waves transmit energy and carry information.

Entranced by the strangeness and beauty of fossil replicas, biology students learn the evolutionary descent of modern horses and how fossils of skulls and teeth bear witness to the history of western North America's climate. The story reveals the visual underpinnings essential to paleontology and geology and how "Aesthetic Expression" (part III) framed Charles Lyell's insights and Charles Darwin's theorizing.

Alfred North Whitehead's dictum, "Understanding is useful," sets the stage for examining the significance of "Conceptual Coherence" (part IV). In the context of studying "earth stories," graduate students in Beijing "mess about" with ice, freshwater, saltwater, and alcohol to explore buoyancy, then apply this concept to the interpretation of tectonic plate motion in a course, "Teach Less/Learn More," crafted in keeping with Joseph Novak's theory of meaningful learning.

"Community Purpose" (part V) threads through all aspects of an elementary school. Following the study of inventions and simple machines, children craft mechanical action figures. A crank turns and Robin Hood teeters on a log, defending himself from the blows of Little John's staff. To build community, children celebrate each other's creations. Community purpose forges collaboration among a school, its community, and a nearby museum, a partnership brokered by a university and a high-tech start-up. A balloon launched from school ascends twenty miles into the sky to record atmospheric conditions. Adolescents and their families retrieve the balloon and with this and other data gathered by backyard sensors, they develop weather forecasts tailored to their region.

By means of playful exploration, artistic expression, conceptual coherence, and community purpose science teaching proceeds on beyond standards. As the sciences abandon the quest for certainty and universality, educators must learn to seize from a dizzying array of diverse disciplines the resources needed to serve timely, local, concrete, and particular community goals.

NOTES

1. Nir Orion and Charles R. Ault, Jr., "Learning Earth Sciences," in *The Handbook of Research on Science Education*, eds. Sandra A. Abell and Norman G. Lederman (Mahwah, NJ: Lawrence Erlbaum, 2007), 653.

2. John L. Rudolph, "An Inconvenient Truth about Science Education," *Teachers College Record* (February 9, 2007), accessed February 27, 2020, https://www.tcrecord.org/Content.asp?ContentID=13216.

3. David Hawkins, *The Informed Vision: Essays on Learning and Human Nature* (New York: Algora Publishing, 2002), 12.

4. Joseph J. Schwab, "The Concept of the Structure of a Discipline," *The Educational Record* 43 (1962): 92.

5. NGSS Lead States, *Next Generation Science Standards: For States, by States* (Washington, DC: National Academies Press, 2013). For current information about the NGSS, visit http://www.nextgenscience.org/.

Part I

PARADIGM LOST

Chapter 1

Descartes' Dream

Frustrated by the tumult of the Thirty Years' War in seventeenth-century Europe—a series of conflicts that decimated populations, especially in Germany—the philosopher René Descartes dedicated himself to placing reason upon a solid foundation. What, wondered Descartes, given the violence attributed to irreconcilable Catholic and Protestant dogmas, might be rational and independent grounds for compelling agreement among people of good will? His answers precipitated the Enlightenment; his followers placed their faith in reason.

Whether by faith or reason, peoples in ancient times and modern have dreamt of joining earthly society and heavenly perfection together in harmony. The Greek word "cosmos" refers to the order of the natural world; "polis" refers to the human community, thought to flourish best when aligned with the nature of the cosmos. Analogy between the order of the solar system and French society, for instance, once justified its rigid, social hierarchy. Picture Louis XIV, the Sun King, seated on his throne, his subjects, in their "fixed orbits." In the present, ecological metaphors express how to improve society, and "natural" often means good.

The European Enlightenment, a particular set of ideas linking cosmos and polis, has shaped the order of American society. Enlightenment idealists, seeking to achieve social harmony through reason, simultaneously birthed scientific rationality, a hallmark of modernity. The idealization of reason was not simply a triumph of progress over backwardness but a response to historical circumstances—centuries of religious war.

THE QUEST FOR CERTAINTY

Descartes' rational ideal has stood through time as the preeminent example of thinking about thinking standing independent of historical and social contexts. "I think, therefore I am" presumably applies to everyone, everywhere, forever. His notions of axiomatic reasoning and geometric coordinate systems (convertible to algebraic expressions) undergird a great deal of scientific apparatus—independent of time and culture. Mathematical abstraction became his edifice of truth.

Descartes rigorously and ruthlessly subjected belief and premise to doubt and skepticism, searching for an irreducible residue as a starting point for his method of reasoning. Descartes' methods and their social value, however, cannot be separated from the time in which he lived. Those times—their political and social conflicts—set his agenda. Philosopher Stephen Toulmin characterizes Descartes' agenda as a "quest for certainty" that united the sciences—until it didn't.[1] Times have changed. The Cartesian paradigm of scientific rationality, born from skeptically winnowing through accepted truths, merits revisiting even if the role of skepticism remains intact.

Through the 1950s, compelling histories of the origin and progress of the quest for certainty, as shared among modern sciences, guided the agenda of school science.[2] In 1962, however, recognition of "paradigm shifts" burst into philosophical view and the quest for "one method to rule them all" (apologies to Tolkien) began to cede its empire.

Scientific Revolution

Paradigms have limited validity and are neither timeless nor universal. They can be generative of fruitful inquiry, yet later expire into irrelevance. Thomas Kuhn's conception of *The Structure of Scientific Revolutions* (editions span 1962 to 2012) "substitutes for the dream of a singular method, applied across the board, the fact of plural explanatory methods, each of which is limited in scope and lifetime."[3] The same as Descartes, Kuhn's thinking emerged in response to an historical and political crisis—the decades of war at the heart of the twentieth century.

Kuhn's thesis arrived on the science education scene at an auspicious moment: the ascendency of science to the commanding heights of national security, with federal funding its primary support. Extrapolating that scientific methods to everyday life diluted their character, scientists wanted political leaders as well as common citizens to understand that their methods of producing and generating knowledge worked in specific ways and in particular contexts. Harvard's president, James B. Conant (a chemist by training), argued that "nothing is further from the procedure of the scientist than a rigorous tabular progression through the supposed 'steps' of the scientific method."[4]

Naomi Oreskes points out that in the decade before Kuhn's *Structure of Scientific Revolutions* began to resonate among philosophers, historians, and sociologists of the sciences, Kuhn expressed a softer stance on the phenomenon of a paradigm shift. He drew an analogy between a bend in the road and a fundamental break in the continuity of thought within a discipline. The perspective on the ground from either direction fails to see around the bend—it looks like an abrupt end or a new start.

From above, the road just bends to head in a different direction. In retrospect, a journey along the road has many stops, the intermediate stages in the evolution of competing conceptions. The beginning and end of the journey may be extremely far apart—"incommensurably" distant conceptually and methodologically.[5] From the perspective of time, the nature of a particular science may seem to evolve more than abruptly shift.

Kuhn's analysis shifted the philosophy of science paradigm itself. The shift set aside dreams of a *universal*, scientific recipe for generating objective truths. The consequence of this shift was to focus on the demands characteristic of particular problems, the novel challenges of new fields of inquiry, and the development of new techniques to create and interpret experiences previously beyond reach. In a nutshell, timely, local, and particular sciences gained stature equal to timeless and universal ones.[6]

The pre-Kuhnian quest to unify disciplines ignores basic insights into how work gets done in laboratories, natural settings, and loosely allied (and competing) research communities. Respected scholars since Kuhn have noted the disparities encountered when crossing the cultural borders from one enterprising, knowledge-producing community to another, finding that theoretical structures and grounds for accepting evidence vary, often incommensurably.[7, 8]

The Case of Geology

Whether through revolution or by evolution, paradigms for the conduct of inquiry have shifted dramatically, often making the explanatory methods of the old incommensurable with those of the new. Although dependent upon physics, the geosciences and climate sciences must, for instance, adapt empirical methods of investigation to complex system modeling tasks contingent upon historical data.

In essence, Kuhn's phrase, "shifting paradigms," refers to changes in patterns of explanation. For example, for many decades the geoscience community, especially in North America, fervently rejected Alfred Wegener's theory of continental drift. Geologists who presumed the immobility of the earth's crust proposed explanations of mountain building based upon cycles of sediment deposition, subsidence, and uplift parallel to continental margins.

Robert Muir Wood's *The Dark Side of the Earth* examines pre-plate theory paradigms in geologic history with acknowledgment of the incommensurability of theoretical commitments and standards of evidence. Grand theorizers suggested in the early nineteenth century that the contraction of the earth on a global scale created the structures of the crust. Influenced by theory in crystallography, they imagined forces tending to compress the spherical earth into a regular icosahedron (a Platonic solid with twenty faces). Picture the earth in consequence as a shriveled apple.

From 1912 to 1960 continental landmass "fixists" antagonized "mobilists" drawn to Alfred Wegener's theory of continental drift by the seductive fit of coastlines (more precisely, continental shelves) across the Atlantic. Ideas of an expanding earth supplanted those of a contracting one in a further effort to account for seams of volcanism.[9]

The fact of seafloor spreading inferred from data obtained in the hunt for Nazi submarines during the Second World War stimulated new thinking about geologic change. Careful mapping of ocean bottoms had revealed mid-ocean mountain chains, rift valleys, and deep trenches associated with chains of volcanoes. Most importantly, sensitive magnetometers detected a pattern of magnetic stripes, symmetrically aligned to zones of rifting and indicative of increasing age of seafloor lavas with distance from the rifting.

These facts established mid-ocean rises as "spreading zones" that produced new oceanic crust and "trenches" as places where seafloor plunged into the mantle. Plate tectonics accounts rather impressively for what happens when continents collide and ocean floor encroaches upon land. The theory makes sense as well of the global distribution of volcanism and earthquakes.[10]

PIAGETIAN PARADIGMS

Piaget argued that the child's development recapitulated the history of advances in human intellect. The child enacts these advances in progressing from sensorimotor reasoning to formal logical operations. Piaget viewed logical reasoning as an extension of biological adaptation—an advanced stage of development, the adult mind, derived from its immature stages.[11]

The Piagetian paradigm, anchored to pre-Kuhnian philosophy of science, reinforces notions of universal aspects of reason and influenced generations of science teachers to think in these terms. In a sense, Piaget's theory is the *Archaeopteryx* of child development theory. It has wings but does not fly well because the anatomy of his theory is tied to a universalist conception of reason.

In Piagetian terms, the child's mind incorporates new information just fine until novelty defies assimilation. Disequilibrium sets in as the mind adapts its logical structure to accommodate novelty. As a consequence, the child's reasoning capacity advances—ultimately, to thinking with mathematical abstraction and deductive logic. Piagetian ideas echo Kuhn's notion of periods of "stable" science disrupted by "fluid" ones as a paradigm shifts. Putting the two paradigms together in science education circles became the task of conceptual change theory—a dominant paradigm in science teaching at the end of the twentieth century.[12]

The first conceptual change researchers studied human learning in the context of the science with the most compelling claim of universality: the physics of motion. The team, led by George Posner, listened to college physics students explain their thinking as they first solved problems of motion and acceleration with Newtonian laws. They next listened to them attempting to solve problems calling for the application of Einstein's relativity theory to interpret particle accelerator data. Students who tended to reinterpret relativity with notions of absolute time and fixed length were said to "Newtonize" relativity.

The Newtonizing students in the study had not "accommodated" the new idea—"accommodation" being Piaget's mechanism for effecting a transition from one stage to the next in cognitive development. Posner's subjects had shifted their conceptual paradigms incompletely. According to conceptual change theory, accommodating a new concept requires first that a student sense an *anomaly* or incongruity in experience and be dissatisfied with current knowledge as a result. A new idea needs to be made *intelligible*—perhaps by being drawn from a familiar context, perhaps by means of analogy.

Next, the new concept must be seen as resolving the tension stemming from awareness of anomaly and dissatisfaction with current knowledge in a *plausible* fashion. Finally, the new concept must prove itself *fruitful* of further inquiry. These criteria remain as valuable guideposts for science teachers—a chimera among educational paradigms that tips its hat to Thomas Kuhn and Jean Piaget.

Conceptual change theory and studies of student misconceptions in science spawned a veritable cottage industry of studies that moved away from the Piagetian notion of reasoning. Science educators have wondered whether student thinking and its progress might mirror conceptual revolutions in the history of science; for example, the historical case of puzzling out how photosynthesis works and the challenge to students to think of the bulk of a tree being built, in effect, from air (the source of carbon atoms).[13] In 1991, Edward Smith proposed "usefulness in a social context" as a fifth dimension influencing conceptual change.[14]

AN IMPOVERISHED VISION

In the twentieth century, the promise of scientific method became an intellectual commodity with unlimited market potential, no matter the problem. Alas, no such generic elixir seems to exist.[15] As domination by "the" scientific method and its successors recedes, the classroom door opens to a fascinating array of topics, each offering value and an interesting take on reality.

Naive learners perceive the history of science as the accumulation of facts and discoveries organized logically and grounded in supporting and improving evidence. This perception disregards how ways of imagining the world are crafted to achieve particular goals embedded in historical contexts. Mischaracterization avoids the hard questions: "What are we to do with our science? On whose authority?" Just turning the crank of "the" scientific method a sufficient number of times to arrive at truth ignores the relation of knowledge to power.

Intuitively, many students and their parents harbor doubts about the legend of progress. Science and weaponry ride in the same boat; consumer goods often carry unsuspected risks. The dangers of nuclear annihilation, environmental contamination, rapid extinction rates, oppressive surveillance, and catastrophic climate change compete for prominence on the horizon of the future. Many know that science's history includes eugenics, scientific racism, and careless disposal of toxic wastes and that these injustices accrue disproportionately to people living in poverty. The dark side of science is inescapable—and intimately connected to public policy.

The embrace of standards and prominence given to cross-national comparisons of student achievement on narrow, technical measures are "theoretically impoverished and politically visionless," claims William F. Kyle Jr.[16] Predominating measures for comparing achievement build upon the myth of steady, incremental, value-neutral, scientific, progress contributing to the betterment of life. This myth leaves out valid social and political contexts and reinforces the impoverished vision of science in the singular.

Taking action to improve communities counts on trustworthy scientific knowledge produced appropriately and adjudicated thoroughly. Neither naive trust nor cynical doubt will do; either invites deception. Wrongly informed action brings suffering. Society truly depends upon public awareness of when and why to place their trust in one or another science. Schooling's historical reinforcement of an image of science as "singular," presently standardized as three dimensions, oversimplifies and misrepresents how science gets done. Repeated misrepresentation results in a public distrustful of sciences that fail to conform to stereotype.

As twenty-first-century science confronts domains that profoundly impact daily living and societal sustainability—medicine and genetics, communication and information technology, energy and climate, pollution and extinction, pandemic and public health—science teaching risks becoming increasingly out of step with science. Continuing to impose on the sciences the quest for universal criteria of reasonableness risks "making our ideas and institutions . . . sclerotic" and resistant to modification in response to "fresh demands of novel situations."[17] By analogy, somnambulantly following Descartes' dream risks making school science obsolete and unable to prepare citizens for the demands of coming decades.

NOTES

1. Stephen Toulmin, *Cosmopolis: The Hidden Agenda of Modernity* (New York: The Free Press/Macmillan, 1990).

2. Ibid., 8. "Newton completed intellectual tasks that were framed in scientific terms by Galileo, and as methodological issues by Descartes, in the 1630s, where Modernity begins for many purposes. For example, British and American universities begin their courses on modern philosophy with Descartes' *Meditations* and *Discourse on Method* while their courses on the history of science present Galileo as the founder of modern science."

3. Ibid., 84. In *Human Understanding: The Collective Use and Evolution of Concepts* (Princeton, NJ: Princeton University Press, 1972), Toulmin presents his case for softening Kuhn's notion of "revolution" to "evolution." The "grasp, growth, and worth" of concepts does happen within individual minds working within communities undergoing paradigm shifts, even if many members of the community resist change to the end of their lives. As expressed in *Human Understanding*, Toulmin's epistemology undergirds much of "Conceptual Change Theory."

4. John L. Rudolph, *How We Teach Science, What's Changed, and Why It Matters* (Cambridge, MA: Harvard University Press, 2019). Rudolph describes how efforts to improve science teaching have tracked moral and value shifts in American political history. Chapter 6, "The War on Method," details the intellectual assault on "the scientific method" as a generic problem-solving strategy. Rudolph's chapter draws from *General Education in a Free Society* (Cambridge, MA: Harvard University Press, 1945), authored by a committee led by chemist and Harvard University president James B. Conant. Through the 1940s and 1950s, Conant targeted public misunderstanding of scientific practice, particularly the idea of a singular, step-by-step method. Conant (and his committee) proposed as a solution the study of cases crucial to scientific advances. Kuhn worked on this agenda.

5. Naomi Oreskes, *Why Trust Science?* (Princeton, NJ: Princeton University Press, 2019), 42. Within a section on the emergence of the field of "science studies," Oreskes quotes and analyzes the "bend" metaphor from Thomas S. Kuhn, *The Copernican Revolution: Planetary Astronomy in the Development of Western Thought* (Cambridge, MA: Harvard University Press, 1957/1992), 182.

6. Toulmin, *Cosmopolis*, 154, 183.

7. Karin Knorr Cetina, *Epistemic Cultures: How the Sciences Make Knowledge* (Cambridge, MA: Harvard University Press, 1999).

8. Peter Galison and David J. Stump, *The Disunity of Science: Boundaries, Contexts, and Power* (Stanford, CA: Stanford University Press, 1999).

9. Robert Muir Wood, *The Dark Side of the Earth* (London: George Allen & Unwin, 1985).

10. In framing of the history of plate tectonics theory, Homer E. LeGrand's *Drifting Continents and Shifting Theories* (Cambridge: Cambridge University Press, 1988) invokes Kuhn's notion of paradigm shifts.

11. Ault, Jr., *Challenging Science Standards: A Skeptical Critique of the Quest for Unity* (Lanham, MD: Rowman & Littlefield, 2015), 136.

12. George J. Posner, Kenneth A. Strike, Peter W. Hewson, and William A. Gertzog, "Accommodation of a Scientific Conception: Toward a Theory of Conceptual Change," *Science Education* 66, no. 2 (1982): 211–227.

13. James H. Wandersee, "Can the History of Science Help Science Educators Anticipate Students' Misconceptions?" *Journal of Research in Science Teaching* 23 (October 1986): 581–597. https://doi.org/10.1002/tea.3660230703.

14. Edward L. Smith, "A Conceptual Change Model of Learning Science," in *The Psychology of Learning Science*, eds. Shawn M. Glynn, Russell H. Yeany, and Bruce. K. Britton (Hillsdale, NJ: Lawrence Erlbaum, 1991), 43–63.

15. John L. Rudolph, "The War on Method," in *How We Teach Science* (Cambridge, MA: Harvard University Press, 2019): 118–138. Rudolph provides a detailed history of mid-century rejection of a singular, five-step, generally applicable scientific method by leading intellectual figures and scientists. These arguments turned attention to the approaches to research undertaken by natural scientists and admonished educators for oversimplifying John Dewey's thought.

16. William C. Kyle Jr., "The Road from Rio to Johannesburg: Where Are the Footpaths to/from Science Education?" *International Journal of Science and Mathematics Education* 4 (2006): 12.

17. Toulmin, *Cosmopolis*, 184.

Chapter 2

Complicated Truths

"How did the Andes become so narrow and tall?" "Lots of earthquakes," accompanied by volcanoes, tsunamis, and landslides, is the simplest answer. The risk of these events makes the study of Andean mountain building compelling to Chilean seventh graders. For them, tangible threats to the regional community elevate concern for the nature of geologic inquiry.

In the dynamic landscape of western South America, horizontal compression has had dramatic shortening and lifting effects. To understand the mechanisms of deformation, the seventh graders placed layers of sand alternating between black and white in clear plastic boxes. To model the rise of the Andes, they turned a crank to move a baffle and compress the layers of sand. Arches and troughs, then faults and overthrusts, took form. Surface patterns reflected the structures below.

They paused the exercise several times to draw the evolving cross-sectional profile and predict what might happen to the structures and the pattern at the surface as compression continued. The patterns from both top and side views resembled what happens at a continental scale.[1]

GEOLOGIC INQUIRY

Geologic insights and arguments rely upon analogical reasoning, often invoked to resolve an anomaly. A proper analogue may advance the interpretation of a particular region's geologic history. In order to explain the formation of the Rocky Mountains, geophysicists Teresa E. Jordan and Richard W.

Allmendinger, for example, drew upon an analogy between the Rockies of North America and a range in northwest Argentina, the Sierras Pampeanas.

The South American case represents a zone of ongoing deformation. The analogous processes in the Rockies happened over 100 million years ago. From their comparison, Jordan and Allmendinger concluded that horizontal compression, rather than vertical stresses, once deformed the Rocky Mountain province.[2]

Chilean earth science teachers are familiar with a gap in volcanism along the Andean coast in the north central region of their country. The gap exists to the west of the Sierras Pampeanas and its cause is the same mechanism responsible for their formation—the horizontal subduction of a tectonic plate.

The process of "flat-slab subduction" (ocean floor remaining in a subhorizontal state as it sinks beneath and rubs against continental rock) resolves the anomaly and accounts for mountain building farther inland. The low angle prevents the conditions required to form magma chambers in line with the Andean volcanic arcs to the north and south of the gap. The explanation has value on two levels. It solves a geologic puzzle and illustrates the geologic approach to figuring out earth history.

ACCENTUATING PLURALITY

In 1869, Thomas H. Huxley (of "Darwin's Bulldog" fame) argued, "No boy nor girl should leave school without possessing a grasp of the general character of science, and without having been disciplined, more or less, in the methods of all sciences."[3] Accepting, as did Huxley, that "the methods of all sciences are identical"[4] (or at least subsumed by the same principles), generations of science teachers have succumbed to the temptation to view science primarily through the abstract lens of singular method.[5]

Valuing "the general character of science" and "the methods of all sciences" fails the task of helping students appreciate the "messy, mottled world" they actually inhabit.[6] Whether casting "the" scientific method from one decade to the next as "habits of mind, practices of science, the nature of science, scientific processes, or inquiry science," the struggle to unify the sciences has faced an insurmountable obstacle: diversity reigns.

Fields differ because challenges differ, and the "nature of science" is best expressed as the "natures of the sciences." After all, concludes Stephen

Toulmin, "tolerating plurality, ambiguity, and lack of certainty is no error, let alone a sin . . . it is part of the price we inevitably pay for being human beings, and not gods."[7]

TOLERATING PLURALITY AND AMBIGUITY

Joseph J. Schwab, perhaps the leading light for teaching science through inquiry, invented a perplexing phrase, "the structure of a discipline."[8] His imagery of "structure," in effect, instantiates how knowledge changes when a paradigm shifts. "Structures" are what construction creates. "Structures" assign meaning to facts, suggest what counts as evidence or not, and provide the bases for arguments warranted within a particular context, as in interpreting the geology of the Rockies by analogy to the Sierras Pampeanas.

Schwab's insight certainly vexed his students. "Tell me," said one, "What is this course about?" Schwab, impressed by his own "clarity and brevity," proceeded to answer—*for twelve minutes*. "Thank you," said the student. "And if the truth is that complicated, I am not interested." Schwab persisted in his venture and for generations educators have struggled to grasp "the sense in which the content of a discipline is warranted and meaningful."[9]

Structure and Style

Kuhn's, Toulmin's, and Schwab's attention to the importance of disciplinary "structures" increased awareness of how historical methodology departed from the stereotype of experimental method. David Kitts, for example, argued in 1977 that geology makes "singular statements" about the past—he called them "retrodictions"—and cannot be expected to construct principles analogous to the laws of physics.[10]

Stephen Jay Gould, drawing upon patterns of extinction in paleontological data, introduced to evolutionary biology the idea of "punctuated equilibrium." Rather than dismissing the fossil records of species changes as recorded throughout earth history as woefully incomplete, Gould's new paradigm accepted the record as an accurate reflection of evolutionary process. He argued that species change occurs relatively rapidly on evolution's timescale, then a long period of "stasis" commonly sets in. His characterization of the "mode and tempo" of evolution ran counter to the "gradualist" paradigm in evolutionary biology.

Gould concluded that the geosciences, paleontology, and important domains of evolutionary biology—and even aspects of cosmology—practiced a "style" of science irreducible to experimental method. These sciences still, of course, sought to associate causes with effects. The methods for doing so had to adapt to the nature of the case. Such "historical sciences," in grappling with time on vast scales, demanded their own scientific paradigm. "Historical," Gould concluded, was science's "second style"—experimental being its first.[11]

Historical sciences depend upon criteria such as "integration across scales" in place of experimental verification.[12] When explanations for the geologic patterns found in crystals, landforms, rock strata, and dating align, confidence in a hypothesis claiming a common cause for these phenomena grows. The classic example of this style of reasoning is the extraterrestrial impact account of extinctions at the end of the Cretaceous Period. The impact hypothesis unites the microscale, shocked quartz at the probable impact site; the mesoscale, the Chicxulub crater structure in the Yucatán Peninsula of Mexico; and the macroscale, the global distribution of iridium-enriched clay sediment, first observed in Gubbio, Italy.[13]

Rationality and Community

The rationality of science resides neither in a universal method nor in an individual mind; the science is in the debate. The stature of the scientific community depends upon its ability to winnow out irresponsible science. The biases and passions, the range of expertise, the diverse identities, the shared commitments, the varieties of perspectives, and the integrity of communications bind research communities. These ligaments consist of checks and balances—argument and review, replication, and scrutiny—essential to achieving consensus or at least providing a compelling basis for widespread agreement. It takes community structure to make the science trustworthy.[14] Clever together is better.

Toulmin notes in his case study of Descartes that post-Kuhnian scholars have gone much further than examining paradigm shifts *within* a science, as exemplified by Gould in biology, Kitts in geology, and, of course, Kuhn in physics. A more comprehensive agenda considers how *external* social, historical, and political factors have influenced problem choices and explanatory patterns.[15] Social milieus, replete with community values, shape scientific agendas.

Science takes pride in being the dominant exemplar of rational thinking and the principle driver of technological progress. Philosopher Paul Feyerabend questions the value of such dominance and claims that the needs of citizens and their communities—sustainable resource utilization and equitable living standards, for example—ought to subordinate scientific enterprises.[16] Understandings of the natural world believed true matter to society because they influence the choice of tools expected to change it and the selection of ideas for teaching the next generation how to do so.

Case and Context

In the wake of Kuhn's influence, an Aristotelian emphasis on "the nature of the case" has displaced Platonic idealism—the idea of a deeper reality hidden by appearances. From Feyerabend's point of view, no rule set or prescription exists (or perhaps ever will exist) to decide the truth of a scientific theory.[17] The case-by-case search for truths eschews the universalism of the paradigm Descartes founded in favor of accepting the grounds and limitations of claims within well-defined contexts. Methods anchored to reflecting upon concrete experience, as opposed to searching for universals, have gained stature. In Stephen Toulmin's words:

> Since the 1960s, both philosophy and science are back in the intellectual postures of the last generation before Descartes. In natural science, the imperial dominion of physics over all other fields has come to an end: ecologists and anthropologists can now look astronomers and physicists straight in the eye.[18]

Grafting patterns of explanation onto carefully crafted experiences is no simple task. Capable educators encourage active investigations in response to essential questions, as in the case of Chilean earth science instruction. Done well, this approach entails complex demands on a teacher's knowledge and skills. When done well, students construct knowledge from experience *and* an understanding of the nature of inquiry within a valued context.

NOTES

1. Claudia Vergara-Díaz, Karla Bustamante, Luisa Pinto, and Hernán Cofré, "Exploring Chilean Seventh-Grade Students' Conceptions of Earth Dynamics Before

and After Model- and Inquiry-based Instruction," *Journal of Geoscience Education* 68, no. 4 (February 21, 2020): 360–370, accessed October 20, 2020, https://doi.org/10.1080/10899995.2020.1725406.

2. Teresa E. Jordan and Richard W. Allmendinger, "The Sierras Pampeanas of Argentina; a Modern Analogue of Rocky Mountain Foreland Deformation," *American Journal of Science* 286, no. 10 (December 1, 1986): 737–764, accessed October 20, 2020, https://doi.org/10.2475/ajs.286.10.737.

3. Thomas H. Huxley, "Scientific Education: Notes of an After-Dinner Speech," 1869 lecture in *Science and Education*, ed. T. H. Huxley (New York: Collier and Son, 1901), 119.

4. Thomas H. Huxley, "On the Educational Value of the Natural History Sciences," 1854 lecture in T. H. Huxley, *Science and Education* (New York: Collier and Son, 1901): 51–52.

5. Charles R. Ault Jr. and Jeffery Dodick, "Tracking the Footprints Puzzle: The Problematical Persistence of Science-as-Process in Teaching the Nature and Culture of Science," *Science Education* 94, no. 6 (2010): 1092–1122.

6. Nancy Cartwright, *The Dappled World: A Study of the Boundaries of Science* (Cambridge: Cambridge University Press, 1999), 6.

7. Stephen Toulmin, *Cosmopolis: The Hidden Agenda of Modernity* (New York: The Free Press/Macmillan, 1990), 30.

8. Joseph J. Schwab, "The Concept of the Structure of a Discipline," *The Educational Record* 43 (1962): 197–205.

9. Ibid., 197.

10. David B. Kitts, *The Structure of Geology* (Dallas, TX: Southern Methodist University Press, 1977); Elizabeth A. Lloyd's *The Structure and Confirmation of Evolutionary Theory* (Princeton, NJ, 1994), Stephen Jay Gould's *The Structure of Evolutionary Theory* (Cambridge, MA: Belknap/Harvard Press, 2002), Kitts argued that geology makes singular statements about the past—"retrodictions"—and cannot be expected to construct principles analogous to the laws of physics.

11. Stephen Jay Gould, S. J., "Evolution and the Triumph of Homology, or Why History Matters," *American Scientist* 74 (1986): 60–69.

12. Charles R. Ault Jr., "Criteria of Excellence for Geological Inquiry: The Necessity of Ambiguity," *Journal of Research in Science Teaching* 35, no. 2 (1998): 189–212.

13. James L. Powell, *Night Comes to the Cretaceous: Dinosaur Extinction and the Transformation of Modern Geology* (New York: W. H. Freeman & Co., 1998).

14. Naomi Oreskes, *Why Trust Science?* (Princeton, NJ: Princeton University Press, 2019).

15. Ibid., 132.

16. Paul Feyerabend, *Farewell to Reason* (London: Verso/New Left Books, 1987).

17. Brian Martin, Heidi Kass, and Wytze Brouwer, "Authentic Science: A Diversity of Meanings," *Science Education* 74, no. 5 (1990): 545.

18. Toulmin, *Cosmopolis*, 168.

Chapter 3

Discipline Diversity

Imagine students in Miami, Florida, graduating with a good understanding of coral reef ecology and the threats to the ocean's reefs globally. Think about their counterparts in Madison, Wisconsin, graduating on the same day, but without knowledge of reef ecology. In its place, however, they have a good understanding of the relationship between soils and climate, with insights into the risks to agriculture of declines in soil moisture in the Midwest as the world warms. Each example underscores learning sciences of importance for solving problems felt significant by the local community.

With respect to efforts to raise the ranking of U.S. students on international comparisons of achievement in science, this scenario looms as nothing less than a nightmare. Acceptance of learning different sciences in different places in different ways for different social (and political) reasons ignores the value of being able to compare science achievement region to region, nation to nation.

Overly burdened educators, sea-tossed by reforms, mandates, class size, and limited resources—and devoted first and foremost to achieving equity through schooling—rightly recoil from opening the Pandora's box of scientific diversity. Hasn't the focus on practices of inquiry, a much-needed remodel of the scientific method, supposedly made the authentic processes of the sciences accessible and useful to all while constraining what teachers are expected to teach? Many practitioners believe so—hence, the momentum of the standards movement.

STANDARDIZATION'S DEFICIENCIES

The lists, matrices, and hierarchies of standards documents guide educators in their quest for a manageable set of goals for coordinating instructions across grade levels and subjects. In the Next Generation Science Standards these goals are framed within three "dimensions": scientific (and engineering) practices, crosscutting concepts, and disciplinary ideas.[1]

Attention to "practices" respects diversity among the disciplines, but this dimension still encourages teachers to think about practices within the limits of eight, generic categories, descended with modest modification, from previous lists of processes. The practices fail to address playfulness with both things and thoughts, or the role aesthetic expression plays in understanding science concepts. The matrices and searchable databases of standards by dimension lack coherence. Grade-level hierarchy is not a substitute, though "learning progressions" are a start.

For student understanding to flourish, creative teachers know they must reach beyond the standards and provide their students with coherent and purposeful contexts for learning science, sometimes in playful and artistic ways. Such ideals promise to enliven the science classroom and escape entrapment by the standards' implicit dictum, "If it does not align, you must not spend the time."

The standards are equally applicable to schools anywhere and everywhere. Yet, in the end, they are emotionless abstractions, itemized generalizations—a set of bureaucratic tools, not a foundation for empowering care for communities. The standards fall short of helping students *feel* the significance of diverse sciences in timely, local, and particular circumstances. By contrast, deeply embedding science teaching in timely, local, and particular circumstances fosters citizen attachment to the landscapes and communities that nurture them.[2, 3]

Clearly, educators may not blithely dismiss adopted standards for teaching the concepts and practices uniting the sciences. Fortunately, standards can be bent and twisted. Or adapted, extended, and, at times, subverted to exploit diversity among the sciences in the pursuit of socially valued outcomes.

ESCAPING THE SINGULARITY PARADIGM

In 1854 Thomas H. Huxley claimed equality between the methods of biology and those of physics.[4] Ever since (and perhaps before), educators have sought to portray the sciences as essentially unified, the peak of separating a one-size-fits-all process of scientific thinking from academic content first being reached in the 1930s.[5] The process approach of the 1960s reinvented this error.

Simply acknowledging that investigators "use a variety of methods," as does the Nature of Science section of the Next Generation Science Standards, fails to break out of the stereotype of science-as-process that, Huxley-like, equates the methods of biology with those of physics. A better insight is a simple aphorism: responses adapt to challenges—to the nature of the case—as needed to deepen understanding in a purposeful context.

The sciences contribute expertise of many kinds to the solutions of important problems. Modeling drought intensity and soil moisture as climate changes in the American Midwest and Central Plains is a good example.[6] Measuring the acidification of the oceans at the end of Cretaceous time serves as another.[7] These very different inquiries recognize the blue planet's ecological melodies and evolutionary symphonies beating to rhythms on different timescales.

Underpinned by plurality and entangled with uncertainty, the sciences must constantly reconfigure to solve twenty-first-century problems. An intellectually honest commitment to this perspective means having students encounter the sciences' diverse responses to distinct challenges.

Grasping the purpose of an inquiry matters to comprehension. Appreciating the demands characteristic of solving particular kinds of problems deepens understanding. These "demands" are not fungible constructs. They are, instead, distinct features of inquiry that truly matter to the development of expertise and the stimulation of interest in a purposeful context, for example:

- Medical research to improve children's health must address the difficulty of obtaining informed consent and the ethics of clinical trials with a vulnerable population.[8]
- Genetic editing to program enzymes to destroy specific strains of bacteria offers remarkable promise but raises questions about the possibility of unintended consequences.[9]

- Reconstructing ancestral human populations as they mix and migrate across tens of millennia takes advantage of methods for sequencing genomes, but quickly becomes entangled in issues of racism.[10]

Speculative and provocative thoughts often emerge when playful exploration of conceptions interacts with rigorous thinking. String theory physicists bear witness to the possible existence of multiple universes.[11] Explaining the origin of matter out of nothing is . . . especially challenging.[12] Perhaps, as imagined by physicist Lee Smolin, black holes birth baby universes, each tuned to different physical constants, a Big Bang their first cry.[13]

Mathematical elegance can be a convincing guide to truth among theoretical physicists, but, unless tested experimentally, not to experimental physicists. The physicist's notion of beauty has historically influenced preferences for one theory over another—new discoveries validating such preferences. Notably, physicist (and blogger) Sabine Hossenfelder asserts that in the twenty-first century this criterion no longer appears to work well.[14]

To Cause or Not to Cause?

In most sciences, not to mention everyday life, nothing seems more real than past and future, cause and effect. Consider, however, particle physicist Sean Carroll's perplexing claim that the "fundamental furniture of reality"—spacetime and quantum fields—cannot be described in terms of cause and effect.[15] "Virtual particles" can become real photons under the right conditions by emerging from, not necessarily nothing, but fields.[16] "And what are fields made of? There isn't any such thing. The fields are the stuff that everything else is made of."[17]

In Einstein's conception of reality, spacetime consists of four dimensions (three linear and one temporal). Past and future, are useful ideas, but questionable guides to reality. The laws of physics treat past and future on an equal footing.[18] In relativity theory, durations are, well, relative. The physics of reality is not only counterintuitive but also rather disconcerting. Reality appears unreal.

Nevertheless, each time people use global positioning satellites to find their way, algorithms derived from special and general relativity combine to pinpoint location. The algorithms correct for the fact that the speedily orbiting satellite's clock runs seven millionths of a second slower per day relative to an earthbound clock (the special relativity correction) and forty-five millionths of a second per day faster in the weak gravitational field high above the earth (general relativity correction). Uncorrected, Global Positioning System errors would be on the order of several miles within a day. We'd be lost without Albert Einstein.[19]

Reality for climate scientists is quite different to that of quantum physicists and general relativity theorists. Extrapolating into the future, in part based upon study of the past, is the primary concern. Climatologists struggle to figure out the causes and effects of global warming. Their explanatory architecture assumes, of course, the physics of fundamental reality. Climate scientists, however, do more than comfortably rearrange the furniture of physics; they construct distinctive edifices of thought and climb their characteristic staircases of method to describe particular aspects of nature.

Cause and effect are clearly central, and, not surprisingly, plagued by uncertainty, especially when inferring long-term climatic effects at local scales. Climate researchers seek to uncover how average weather shifts temporally and geographically. In order to succeed, climate scientists build computer models, marshal paleontological data, and collect a host of observations in real time from ground level (and ocean depths) to high in the atmosphere. They grind this assortment of physical parameters, fossil information, and weather recordings into reliable claims in a statistical mill—ideally, constraining uncertainty in measured ways.

The hybridized approach of climate science generates insights into storm frequency, drought severity, rainfall patterns, and sea level rise on various timescales. Figuring out actual *reasons* for climate change, essential to informing efforts to mitigate its threats, advances the science. Criteria for adjudicating its claims—the fit of the model to observations of actual weather, the completeness of atmospheric data, the distribution in time and space of fossil records—are, as in any science, commensurable with the nature of the case. And, unless willfully ignorant, people trust their claims.

Grand Challenges

The National Science Foundation (NSF) screens proposals for cutting-edge science with ten "big ideas" reflecting "grand challenges."[20] Here are several of the big ideas from the Foundation's current agenda:

- Understanding how genetic and environmental information determines the precise characteristics of organisms of interest to medicine, agriculture, and ecology;
- The future of work at the human-technology frontier as artificial intelligence permeates interactions;
- Opening "windows on the universe" (neutrinos and gravitational waves) that expand ways of observing reality;
- Navigating the Arctic;
- Leading the next quantum revolution;
- Conducting basic research in math, statistics, and computer science to drive discovery through visualization, data mining, and machine learning.

The grand challenges and big ideas setting the NSF's table demand diverse thinking and techniques irreducible to a single method. In response to grand challenges, distinct sciences assemble in novel configurations. No discipline reigns supreme, no single method dominates. Solutions constrain, but do not eliminate uncertainty. Good things happen when the sciences discard the quest for universality.

Current standards for teaching science no longer mythologize "the" scientific method, but they do continue to integrate limited lists of core ideas from just three disciplines with a small number of abstractions. This depiction embodies the lost Cartesian paradigm of science, the quest for certainty (and, by implication, unity)—an intellectual artifact whose pedigree extends back to the seventeenth century.[21]

NOTES

1. NGSS Lead States, *Next Generation Science Standards: For States, by States* (Washington, DC: National Academies Press, 2013). For current information about the NGSS, visit http://www.nextgenscience.org/.

2. David A. Gruenewald and Gregory A. Smith, *Place-Based Education in the Global Age: Local Diversity* (Mahwah, NJ: Lawrence Erlbaum, 2008).

3. Gregory A. Smith and David Sobel, *Place-and-Community-Based Education in Schools* (New York: Routledge, 2010).

4. An interesting window into the foundations of debate about a universal science and its implications for science education can be found in Thomas H. Huxley's 1854

essay, "On the Educational Value of the Natural History Sciences," in *Science and Education*, ed. T. H. Huxley (New York: Collier and Son, 1901), 40–63.

5. John L. Rudolph, *How We Teach Science: What's Changed and Why It Matters* (Cambridge, MA: Harvard University Press, 2019), 116–117.

6. Benjamin I. Cook, Toby R. Ault, and Jason E. Smerdon, "Unprecedented 21st Century Drought Risk in the American Southwest and Central Plains," *Science Advances* 1, no. 1 (February 12, 2015). https://doi.org/10.1126/sciadv.1400082.

7. Michael J. Henehan, Andy Ridgwell, Ellen Thomas, Shuang Zhang, Laia Alegret, Daniela N. Schmidt, James W. B. Rae, James D. Witts, Neil H. Landman, Sarah E. Greene, Brian T. Huber, James R. Super, Noah J. Planavsky, and Pincelli M. Hull, "Rapid Ocean Acidification and Protracted Earth System Recovery Followed the End-Cretaceous Chicxulub Impact," *Proceedings of the National Academy of Sciences* (PNAS) 116, no. 45 (November 5, 2019). https://doi.org/10.1073/pnas.1905989116, accessed November 11, 2019, https://www.pnas.org/content/116/45/22500.

8. Canadian Paediatric Society, "Ethical Issues in Health Research in Children," *Paediatric Childs Health* 13, no. 8 (October 13, 2008): 707–712. https://doi.org/10.1093/pch/13.8.707, accessed November 11, 2019, at https://www.ncbi.nlm.nih.gov/pmc/articles/PMC2606083/.

9. Knuvl Sheikh, "Is Crispr the Next Antibiotic?" *The New York Times* (November 5, 2019): D3, accessed November 11, 2019, https://www.nytimes.com/2019/10/28/health/crispr-genetics-antibiotic-resistance.html.

10. David Reich, *Who We Are and How We Got Here: Ancient DNA and the New Science of the Human Past* (New York: Pantheon Books).

11. Sean Carroll, *The Big Picture: The Origins of Life, Meaning, and the Universe Itself* (Dutton/Penguin Random House, 2016), 306–308.

12. Ethan Siegel, "How Did Matter in Our Universe Start from Nothing?" *Forbes* (January 5, 2018), accessed November 11, 2019, https://www.forbes.com/sites/startswithabang/2018/01/05/how-did-the-matter-in-our-universe-arise-from-nothing/#550220c24c2e.

13. Lee Smolin, *Time Reborn: From the Crisis in Physics to the Future of the Universe* (New York: Houghton Mifflin Harcourt, 2013), 124–125.

14. Sabine Hossenfelder, *Lost in Math: How Beauty Leads Physics Astray* (New York: Basic Books, 2018).

15. Carroll, *The Big Picture*, 29.

16. Charles Q. Choi, "Something from Nothing? A Vacuum Can Yield Flashes of Light," *Scientific American* (February 12, 2013), accessed February 27, 2020, https://www.scientificamerican.com/article/something-from-nothing-vacuum-can-yield-flashes-of-light.

17. Carroll, *The Big Picture*, 173.

18. Ibid., 54–55.

19. Peter Galison, *Einstein's Clocks, Poincaré's Maps: Empires of Time* (New York: Norton, 2003), 288–289.

20. 10 Big Ideas for Future NSF Investments, accessed November 16, 2020, https://www.nsf.gov/about/congress/reports/nsf_big_ideas.pdf.

21. Toulmin, *Cosmopolis*. Toulmin examines the historical circumstances of the origin, growth, and persistence of the quest for certainty and a unified, generalized characterization of scientific method. The rationale of *Science Beyond Standards* hinges on his analysis.

Chapter 4

Menacing Mudflats

The distinguished geophysicist leaned forward, pinched his nose between the thumb and index finger of his left hand, placed his right arm within the pocket formed by his left, swung his shoulders, and raised his forearm at the elbow to imitate an elephant's trunk. Disturbed by infrasound noise, the geophysicist and a small herd of teachers ran up a gentle slope. Realizing an earthquake was imminent, they trumpeted their alarm.[1] Silly people; infrasound is for pachyderms.

This episode of elephant mimicry introduced a summer workshop on seismic hazards in the Pacific Northwest.[2,3] Participants had the good fortune to learn from Brian F. Atwater, the geomorphologist who uncovered the records of Cascadia subduction zone great earthquakes in the 1980s.[4] They drew core samples from the mudflats at low tide in Young's Bay near the mouth of the Columbia River and inspected them carefully, locating deposits of sand sheets that had swept up the estuary multiple times over multiple millennia.

The vertical record was clear. It revealed repeated cycles of rapid burial, then finer sediment accumulation, with zones sometimes supporting Sitka spruce forests. Had the land risen and fallen? Had great floods swept the coast? Had floods come from enormous snowmelt and rainfall swelling rivers or floods from ocean floor earthquakes? If from earthquakes, from where in the Pacific basin? Off Chile, Indonesia, the Aleutians, Japan?

DEAD CEDARS AND JAPANESE CHRONICLES

In years prior, Atwater found stands of dead cedar trees at several sites along the Washington and Oregon coasts and named them "ghost forests" (figure 4.1). At low tide in other locations, he noted the stumps and roots of long-dead Sitka spruce trees emerging from the mudflats (figure 4.2). As quoted in *Inquiry and the National Science Education Standards*,[5] he wondered, "What could have killed so many trees over so wide an area?"

On the other side of the Pacific, dutiful eighteenth-century magistrates in Japan had recorded the arrival of a series of tsunami waves at particular times and specific places along an extensive coast. These arrival times and wave intensity, wound backward, converged on the Cascadia subduction zone of the Pacific Northwest. Atwater and his fellow detectives found the culprit. An earthquake, somewhere far enough away from Japan that shaking gave residents no warning, had produced a destructive tsunami on their shores (an "orphan tsunami")—and a much larger one near the parent

Figure 4.1 The "Ghost Forest of Copalis," a Stand of Dead Cedar Trees on the Banks of the Copalis River in the State of Washington, Casualties of a Great Subduction Zone Earthquake in 1700. *Source*: Photograph by Robert F. Butler. Used with permission.

Figure 4.2 Roots of a Sitka Spruce Tree, a Victim of the 1700 Cascadian Subduction Zone Megathrust Earthquake, Exposed at Low Tide in Willapa Bay, Washington, along the Shore of Long Island. *Source*: Photograph by the author.

quake's epicenter, a great earthquake of magnitude 9 (moment magnitude scale).[6]

The similarity of earthquake scale in the Cascadian zone to the seismic catastrophes in Chile (1960), Alaska (1964), and Indonesia (2004) shocked the geoscience community.[7] "The Really Big One," a 2015 article in *The New Yorker* by the journalist Kathryn Schulz, alarmed the citizenry.[8]

Residents of Portland, Oregon, began to worry about the collapse of multistory school buildings. When, everyone wanted to know, might the big one happen? Though infrequent—spaced centuries apart—the most accurate answer to the worrisome question was "at any time."[9]

The statistical odds are a bit more comforting. The *Oregonian* newspaper updated the probabilities shared with the public to a 15 to 20 percent chance within the next half century.[10] There's probably time for retrofitting and purchasing insurance; building codes have already been changed and emergency services bolstered.

Gumdrop Models

Given the devastation of the 2011 Tōhoku earthquake and tsunami, measurements of crustal deformation measured by a global positioning system monitoring station near Newport, Oregon, raised many eyebrows. Compression shifts the station to the east several millimeters per year, then, curiously, its motion slips slowly back to the west a few millimeters during a "silent" tremor. The data suggest symmetrical straining on the Japanese and North American sides of the Pacific.

The domed monitoring structure looked like a giant gumdrop perched on toothpicks—an unlikely harbinger of doom. The tremors appear as spikes on a seismograph but are barely detectable by people (or elephants). Episodic tremor and slip events bear careful watching. Perhaps acceleration in these "episodic tremor and slip" events signals an oncoming, large, subduction zone earthquake.

Back in the classroom, Portland teachers distribute genuine gumdrops and toothpicks for constructing models of global positioning stations for monitoring episodic tremor and slip. Students move the models on graph paper on a one-to-one scale with the earth in terms of distance but in more rapid time—a few minutes, not years. Alarmingly, the scale of displacement resembles observations of strain in Japan in the decades preceding the devastating Tōhoku events.[11] The exercise brings the message home.

VALUING RESILIENCY

Novel topics and new understandings of the natural world—such as the seismicity of earth's tectonic plate boundaries—challenge the sciences to adapt. Science education rides in the same boat, but with an added responsibility to prepare students for lives of social responsibility in a democratically governed nation. How, when, and with what consequences plates move matters to communities in seismically active regions—and to the school science taught there. Citizens living near an active plate margin share a responsibility for ensuring a resilient response to seismic hazards.

As conceptions and methods prove their worth, they dispel the lingering influences of Descartes' dream and awaken a cornucopia of resources for fulfilling social responsibility. Communities struggle to achieve resiliency in the face of many challenges: racism, poverty, resource depletion, changing demographics, and natural hazards. "Achieving "resiliency" stands as an exemplar of how social value calls upon particular sciences and not the inculcation in a universal notion of science.

STANDARDIZATION DEFICIENCIES

Democratizing intellect as the habits of mind common to all the sciences has become educational apple pie: "the" scientific method its simplest recipe. Commitment to the proposition that all sciences are created equal; that they are endowed by their researchers with the same universal properties; and that among these are observation and inference, classification, the control of variables, and the pursuit of testable prediction will not suffice.

To John Dewey goes much of the credit (or blame) for characterizing science as a generic process. Historian and philosopher of science Michael R. Matthews points to Dewey's 1910 essay, "Science as Subject-Matter and as Method," as the key moment, and to Dewey's assertion, "Scientific method . . . represents the only method of thinking that has proved fruitful in any subject," as the key principle.[12]

In his 1910 essay Dewey laments conventional teaching of science as little more than the accumulation of reliable information and not as the acquisition of a scientific attitude. He thought of science as "a method of thinking." For Dewey, failure to address these dimensions deterred students from pursuing science. Rote learning numbed minds and squashed interest. Authoritarian schooling reinforced dutiful, ductile, and docile behavior.

As John Dewey recognized, excellent science teaching recreates for students the experiences and logic of action undertaken by scientists in their original investigations, field study of tsunami deposits near the Oregon coast being a clear case example.[13] Progressive educators value immersing students in contexts of inquiry. Educating them for lives of responsibility demands more.

Dewey believed civilization's future depended upon teaching "the scientific habit of mind." As the twentieth century unfolded, John Dewey's popularizers distilled his universalist views about scientific methodology into a short list of steps comprising "the" scientific method.

Dewey opposed stultifying regimentation and its promotion of rote learning, as do many educators today. However, such reservations have failed to preclude the codification of his thinking into the five (or four, or six, or eight depending on the source) steps of "the" scientific method. Go ahead, Google it. Then compare these steps to Huxley's 1854 list prefaced with the claim that "the methods of all sciences are identical; and these are" as follows:[14]

1. *Observation of facts*—including artificial observation which is called experiment.
2. *General propositions*—tying up similar facts into bundles called *comparison* and *classification.*
3. *Deduction*—making inferences from the general propositions.
4. *Verification*—ascertaining whether an anticipation (inference) is correct.

Dewey's Five Steps

Dewey never presumed to reduce the complementarity of thinking and doing to such a formula, although he came close. John Rudolph has uncovered the history of this chapter of science education. In *How We Think*, Dewey proposed a five-step heuristic linking pragmatic thought to education.[15] Slightly paraphrased, his steps were as follows:

1. Feel the difficulty of a situation.
2. Locate and define this difficulty.
3. Suggest possible solutions.
4. Examine the implications of these solutions.
5. Conduct further observation and experiment leading to a solution's acceptance or rejection.

In following these steps, the learner, believed Dewey, recapitulated the experience of doing science.

Deweyan science works to solve problems. It does so by constructing mental and physical artifacts useful in the conduct of inquiry. Ideas and methods—both engineered—adapt to the demands of the problem. The pluralistic, fluid vein embedded in Deweyan thought, and its depiction of concepts and methods as intelligent adaptations for problem-solving, remains viable and attractive. Ironically, through his pragmatic focus on experience as teacher, Dewey undermined his own notion of scientific method as the "only method of thinking fruitful in any subject." Acknowledging the demands on thinking made by the intrinsic characteristics of diverse experiences opened the gates of scientific inquiry in school science to pluralism.

The Singular-Plural Divide

Dewey's pragmatic attention to the experience of inquiry in a variety of contexts straddles the divide between characterizations of the sciences as singular versus plural. The singular depiction elevates their common and supposedly essential features, while the plural characterization emphasizes understandings of their independence. The pluralist critique exposes the beauty and liabilities of a field of inquiry in light of its own explanatory aspirations. This critique, of course, confuses comparisons among the sciences to determine which ones are better examples of scientific inquiry than others. But it does offer a means to warrant trust—circumscribed by context.

Disunity liberates disciplines from a monolithic commitment to a universal method of truth-finding—making each field of inquiry free to match the demands of its own problems with methods proved fruitful in experience, as opposed to drawn from and aligned with a general theory of rationality.

Does school science mirror this liberation? Or does school science reinforce the Cartesian quest for certainty? Read between the lines, the previous paragraph applies to school science. Is school science compelled to draw from and align with a greatly curtailed, fixed set of abstract categories? Or are schools free to match science teaching to the needs of society as experienced within social, political, historical, and geographical contexts?

Standardization seems to have codified "what to teach" and "to what end" at just the wrong moment in intellectual, social, and scientific history. The conception of science, whether enumerated as processes in the 1960s or practices in the 2010s, has baked Descartes' agenda into school science. Teachers cannot exclaim "Disunity dominates science!" because they have been taught otherwise. Unity dominates the standards—the very sense of scientific unity criticized by Thomas Kuhn and warned about by Stephen Toulmin.

Overly ambitious standardization of "what to teach" risks failing to adapt science curricula to the social and historical contexts that shape the lives of students in particular places and times. Alignment with three misplaced abstractions, independent of these contexts, awkwardly restrains instruction. If schooling is to foster among students a sense of caring for their communities and landscapes—whether facing seismic risks, climate change, or threats to health—the curriculum needs to tap fields of inquiry trusted to contribute solutions to such problems and escape from the shadows cast by alignment with standards.

TIMELY, LOCAL, PARTICULAR

Reformers have organized curricula to achieve socially valued outcomes since the inception of school science. As early as the mid-nineteenth century, Thomas Huxley argued that every boy and girl ought to leave school with a grasp of the "general character of science." John Dewey sincerely believed that civilization's future depended upon citizen appreciation of "the scientific habit of mind," and his perspective dominated twentieth-century reform ideology.

School science rhetoric continues to conform to a conception of inquiry that since 1630 has privileged abstract, timeless, general, and universal theories.[16] The twenty-first century finds the American Association for the Advancement of Science's (AAAS) *Project 2061: Science for All Americans* continuing to promote the "common themes and habits of mind spanning all the disciplines" as essential to designing a "coherent vision of the knowledge and skills [for] every high school graduate."[17] The statement combines Dewey's words with Huxley's.

In contrast, adapted to community needs, curricula to raise "questions of rational adequacy [that] are timely not timeless, concrete not abstract, local not general, particular not universal."[18] Standards purport to focus the curriculum on the most important aspects of science, but no aspect is more important than the value of science to citizens and their communities.

PRECIOUS LIVES

A community confronting the risks of earthquake and tsunami hazards must place a high value on resiliency. Achieving resiliency in a seismically risky place depends upon having adequate knowledge of how to recognize potential hazards, mitigate risks, and prepare for recovery, as exemplified by the case of the Cascadia subduction zone in the Pacific Northwest of the United States. The story exemplifies the value of timely, concrete, purposeful, local, and particular scientific knowledge.

Blindness to, or fateful acceptance of, the inevitability of catastrophe disempowers citizens. Community-level participation and rehearsal in how to prepare for and respond to natural hazards, buttressed by awareness of science made local and timely, boost resiliency.

Gawker Science

In the midst of natural disasters, people lose family members, neighbors, and close friends. The scale of the tragedy overwhelms the community, and economic hardship abounds, especially impacting communities already mired in poverty. Suffering may endure for a long time and recovery remains elusive. Explanations of the causal role played by plate motions and boundaries in the occurrence and severity of earthquakes feel paltry and inadequate to audiences for whom the disaster is a life-altering and soul-scarring experience. Earthquakes and tsunamis are the cause of untold grief. Bhaskar Upadhyay, who survived the earthquake of April 25, 2015, in Nepal, wrote:

> All turned out to be happy for my family but not so happily for thousands of fellow Nepalese. The massive 7.9 quake took away so many precious lives, thousands of years of cultural heritage, peoples' homes, and livelihoods but left behind lifelong psychological and physical trauma. One of the most painful legacies of this earthquake that I witnessed in Nepal was the psychological trauma that children had to endure. Many were getting used to living with the fear of another quake and for many this will become a life-long trauma. . . . Out of 36,000 homes only 5,000 were left standing but even these homes were unsafe to inhabit.[19]

Upadhyay wondered, "Does science matter to the children of Nepal?" He had ventured to Nepal, his birthplace, with the intent of crafting curriculum underscoring how to address the problem of food insecurity in local communities. The destruction he witnessed of temples and monuments in Kathmandu created a feeling of immense sadness, as if he had lost part of his identity. The experience deepened his lifelong commitment to the principle "that without strong connections between students' own experiences and their identities, there is very little value seen in learning science."[20] Dispassionate science, unable to predict the timing and severity of earthquakes—just their inevitability—does little to ameliorate fear.

Upadhyay concludes that the science of seismic hazards must not only promote being safe but also guide physical, emotional, and cultural rebuilding in the event of catastrophe. Unless science finds a way to foster resiliency, ease suffering, and improve security, its role in tragic circumstances reduces to that of just another "gawker,"[21] a person sitting in comfort reading a newsfeed.

Optimistic Response

The people of the Cascadian coast have at their disposal strongly validated scientific knowledge. The historical record of earthquakes and tsunamis, attached to a convincing explanation of their cause, is a trustworthy basis for forecasting dire threats to the future of their communities. How will they respond?

Some may turn away and refuse to think about the danger. Some may look at the video images of Tōhoku and shrug, "It's fate; there's nothing I can do." Others will pack a go bag; bolt their house to its foundation; support funding for "vertical evacuation platforms" accessible to beachcombers; lobby for the state to fund inertial cutoff valves for gas lines to private homes; and voice their concerns in town hall meetings on emergency preparedness, zoning, and building codes.

In coastal Cascadia, envisioning a resilient response to the threat of a tsunami began with teachers, scientists, local citizens, first responders, and park rangers together on a hillside making elephant noises. They charged up the safest evacuation route—as mapped by a high school science class. Particular, local, and timely science mattered to them.

NOTES

1. David Malakoff, "About Can Elephants Predict Disaster?" *PBS Nature* 24, no. 3, written and produced by Jeff Swimmer, narrated by Paul Christie, photography

directed by Bob Poole (New York: Optomen Productions, Inc. and Thirteen/WNET, November 12, 2005), accessed August 12, 2020, https://www.pbs.org/wnet/nature/can-animals-predict-disaster-introduction-2/134/.

2. "Teachers on the Leading Edge (TOTLE): Guiding Northwest Educators from Subduction to Eruption," Robert Butler, project director, accessed August 12, 2020, https://sites.up.edu/totle/. Incorporated Research Institutions for Seismology (IRIS), "TOTLE: Earth Science Teacher Professional Development Featuring Pacific Northwest Earthquake and Tsunami," https://www.iris.edu/gallery3/research/2010proposal/E_and_O/Brian2.

3. Robert F. Butler, Ellen Morris Bishop, Charles R. Ault Jr., Brian Atwater, Bonnie Magura, Christopher Hedeen, Robert Blakely, Roy Wells, Karen Shay, Rebecca Wagner, Travis Southworth-Neumeyer, and Darci Connor, "Linking Middle-School Teachers to EarthScope," *Eos, Transactions, American Geophysical Union* 87, no. 26 (2006): 257, 261.

4. Brian F. Atwater, Musumi-Rokkaku Satoko, Satake Kenje, Tsuji Yoshinobu, Ueda Kazue, and David K. Yamguchi, *The Orphan Tsunami of 1700: Japanese Clues to a Parent Earthquake in North America*, 2nd ed., Professional Paper 1707 of the United States Geological Survey (Seattle: University of Washington Press, 2016).

5. Center for Science, Mathematics, and Engineering Education, National Research Council, *Inquiry and the National Science Education Standards: A Guide for Teaching and Learning* (Washington, DC: National Academy Press, 2000), 2.

6. The Richter magnitude scale is a historic development but is no longer used as a measure of earthquake size. The most commonly used scale is now the "moment magnitude scale." Magnitude 7 earthquakes are referred to as "major" earthquakes whereas magnitude 8 and 9 earthquakes are referred to as "great" earthquakes.

7. Robert F. Butler, "Cascadia Earthquake Science and Hazards," in *Natural Disasters and Risk Communication: Implications for the Cascadia Subduction Zone Megaquake*, eds. C. Vail Fletcher and Jennette Lovejoy (Lanham, MD: Lexington Books, 2008), 15–47.

8. Kathryn Schulz, "The Really Big One," *The New Yorker* (July 20, 2015), accessed May 11, 2020, https://www.newyorker.com/magazine/2015/07/20/the-really-big-one.

9. Chris Goldfinger, C. Hans Nelson, Ann E. Morey, Joel E. Johnson, Jason R. Patton, Eugene B. Karabanov, Julia Gutierrez-Pastor, Andrew T. Eriksson, Eulalia Gracia, Gita Dunhill, Randolph J. Enkin, Audrey Dallimore, and Tracy Vallier. *Turbidite Event History—Methods and Implications for Holocene Paleoseismicity of the Cascadia Subduction Zone*, United States Geologic Survey Professional Paper 1661-F (Reston, VA: United States Geological Survey [USGS] 2012), doi.org/10.3133/pp1661F.

10. Kate Williams, "Risk of Major Quake on Cascadia Subduction Zone Higher than Previously Thought, The Oregonian/Oregonlive.com (August 09, 2016; updated April 22, 2019) accessed August 12, 2020, https://www.oregonlive.com/pacific-northwest-news/2016/08/risk_of_major_quake_on_cascadi.html.

11. Roger Groom, Shelley Olds, Robert F. Butler, and Jenda Johnson, "Episodic Tremor and Slip in a Middle School Classroom? Absolutely!," *The Earth Scientist* 27, no. 1 (2011): 13–19.

12. Michael R. Matthews, "Improving Science Education," in *Time for Science Education: How Teaching the History and Philosophy of Pendulum Motion Can Contribute to Science Literacy* (New York: Kluwer, 2000), 15. Matthews quotes from John Dewey, "Science as Subject-Matter and as Method," *Science* 31 (1910): 122. Reproduced in *Science & Education* 4 (1995): 391–398.

13. Ibid., 137–139.

14. Thomas H. Huxley, "On the Educational Value of the Natural History Sciences," in *Science and Education*, ed. T. H. Huxley (New York: Collier and Son, 1901), 51–52.

15. John Rudolph, "Epistemology for the Masses: The Origins of 'The Scientific Method' in American Schools," *History of Education Quarterly* 45 (2005): 341–376.

16. Stephen Toulmin, *Cosmopolis: The Hidden Agenda of Modernity* (New York: The Free Press/Macmillan, 1990), 35.

17. Jo Ellen Roseman, "Science for All Americans: Timely or Timeless?" *NSTA Reports* 20, no. 7 (2009): 3.

18. Toulmin, *Cosmopolis*, 34.

19. Bhaskar Upadhyay, "Sociology of Earthquake: Does Science Matter to Children of Nepal?" in *Proceedings of Science Education at the Crossroads*, eds. John Settlage and Adam Johnston (Cleveland, OH: October 1–3, 2015), 54, accessed October 13, 2020, https://sciedxroads.org/wp-content/uploads/2014/10/2015_XRoads.pdf.

20. Ibid.

21. Ibid., 55.

Part II

PLAYFUL EXPLORATION

Chapter 5

Wavy Elephants

For several days, the children in the fourth grade had experimented with waves and sounds. Today's science class was underway when a bright flash of lightning struck a tree on the playground. At nearly the same instant a deafening boom shook the classroom. The storm continued, its host of sounds competing with each other: the beating of the rain, the wailing of the wind, the loud claps of thunder. Inaudible rumblings, lower than the loudest and lowest pitch sounds, could be felt coursing through the building, rattling the doors and windows. The air itself seemed to throb against the children's chests.

ELEPHANT COMMUNICATION

Serendipitously, Mr. Link's fourth graders had just watched a video, produced by the Cornell Laboratory of Ornithology, about elephant communication.[1] Now, the skies themselves were speaking in infrasound, the language of elephants in frequencies inaudible to humans.

The frightening storm held everyone's attention. A girl cried and asked to call her mother. A direct instructional approach to teaching another lesson linked to Physical Science Standard #4 could now wait. To calm the class, Mr. Link began to tell a story.

How Elephants Became Able to Speak to Each Other When Far, Far Apart

Long, long ago, the Elephant's Child, ever so full of satiable curtiosity [*sic*], as you may recall from Rudyard Kipling's *Just So Stories*, once found its bulgy boot of a nose clasped in the jaws of a crocodile. As the Elephant's Child pulled, and pulled, and pulled, his nose stretched and stretched and stretched . . . to become a quite serviceable trunk, "a really truly trunk same as all Elephants have to-day."[2]

Unbeknownst to most people, later that same, hot day, the Elephant's Child returned to the shore of the murky Limpopo River, lowered its remarkable new trunk into the water, and started to suck in a refreshing drink. Alas, at that very moment, a Bony Hyoid Fish was swimming by and, yes, up the marvelously prehensile, tubular trunk (a stunning combination of muscular upper lip and flarable nares) it went. Fortunately, drinking with a trunk is a two-stage affair; otherwise the Bony Hyoid Fish might have remained stuck right there. The little elephant quickly squirted the contents of its trunk down its throat. And down, of course, went the Bony Hyoid Fish, only to find a new place to lodge.

The Elephant's Child coughed and coughed, swallowed hard again and again, swiveled its jaws, lashed its tongue all about, and called and called for help. The Bony Hyoid Fish stretched but would not budge. If anything, it lodged even deeper, wriggling and squiggling. Eventually, the Elephant Child's throat began to stretch, a pouchy-pouchy stretch. Even its voice box began to flex and move lower in its throat. At last, by leaning against a large boulder and rapidly self-executing a Heimlich maneuver, the Elephant's Child managed to dislodge the Bony Hyoid Fish, though some of the bones of its fins remained throat-stuck.

"Help me, help me," cried the Elephant's Child. "I have hyoid bones stuck in my throat! I'm here, by the Baobab Tree and the Big Rock on the shore of the Limpopo River!"

Had there been any elephants nearby, they might have heard blasts trumpeting from the Elephant's Child's trunk, a "noise . . . louder than several brass bands."[3] But on that day none were in range of these high pitch vocalizations. Surprisingly, however, even though at great distances, several members of the herd, in fact, detected something unusual: extremely low frequency vibrations (less than 20 Hertz) known to humans as "infrasound." The rumblings were not random at all, but instead patterned in the language of elephants. Listening carefully to this information let them know exactly where to find the Elephant's Child, whose impertinence and 'satiable curtiosity had, despite frightening difficulty, rewarded it again.

In response to all the ruckus, soon a number of his cousins and aunties from far away arrived at the familiar Baobab Tree to offer comfort. They were all greatly surprised, yet enormously pleased, to realize that the Elephant Child's vocalization had traveled so clearly through the dense jungle.

Elephants, of course, know just how effectively dense vegetation absorbs high frequency sounds—such as trumpeting. How wonderful that the little elephant, in its thirst, had solved this problem. Soon all the Elephant Child's cousins, aunts, and uncles—and even its third cousins twice-removed—wanted to speak in such low voices, too. So, one by one they went to the river hoping to swallow a Bony Hyoid Fish. Quickly, the herd spread out to share this wonderful new adaptation with many more of their proboscidean relatives (for this was long ago and there were elephants of different kinds—mastodons, mammoths, gomphotheres, and such).

Elephants thus learned *how to use acoustic energy, propagated as low frequency sound waves, for information transfer*. Ever since, thanks to choking on Bony Hyoid Fish, messaging with infrasound has proven ever-so-popular among elephants of all sorts. And that, O Best Beloved, is how elephants came to be able to talk to each other when far, far apart.

The Evolution of Proboscidean Infrasound Communication Greatly Abridged

The storm subsided; the story ended. How, really, are elephants able to speak in infrasound? Elephant's ability to communicate infrasound utilizes a throat pouch supported by a peculiar arrangement of muscles, ligaments, and five, delicate hyoid bones. Together with a flexible larynx (voice box); long, loose vocal cords; and the large resonating chamber created by the mouth and trunk, the throat pouch is able to send infrasound signals far and wide.[4] The handy pouch also may store water for crossing dry wastelands or spraying water out the trunk to provide relief from heat, as the Elephant's Child so capably demonstrated to his kin.[5]

Darwinian responses among prehistoric populations of proboscideans to climate alterations of habitat evolved in the neighborhood of twenty-five million years ago. They evolved big heads, resonating throats, and a flexible hyoid structure—distinctive features of elephants. The origins of these traits appear to coincide with arid conditions advancing across the changing landscapes at the time marking the boundary of the Oligocene and Miocene epochs. Highly social elephant herds ranged widely to secure food. The ability to produce and decode infrasound proved advantageous because it aided communication among the dispersed members of the heard. Water held in the throat pouch showered hot bodies and quenched thirst. In essence, the several functions of the hyoid apparatus appear to have contributed to proboscidean adaptability to changing habitats. An elephant in distress, or an elephant sharing where to find food and water, became able to share this information by calling to others with infrasound.[6]

SOUND INFORMATION

What is a whimsical story about elephant talk doing in a science class? Sound travels in waves and wave patterns can carry information. A single disturbance or a cycle of disturbances—a vibration, whether of vocal cords or guitar strings—may initiate a wave or set of waves. Resonance chambers in musical instruments amplify sounds; elephants amplify sounds by resonating them in their heads and throats. How might fourth graders send messages in tones audible to humans? By speaking, of course, or perhaps, with simple sound-making devices.

Nowadays, the *Next Generation Science Standards* prompts fourth-grade, human children, especially those full of 'satiable curtiosity, to learn how to communicate using the acoustic transfer of energy in patterned waveforms of varying frequencies (greater than 20 Hertz, of course).[7] They do not have the vocabulary to say it that way, but they certainly speak in audible tones.

The Higgins School District had assigned to Mr. Link and his fellow team of fourth-grade teachers the task of teaching Physical Science Core Idea #4, "Waves and Their Applications in Technologies for Information Transfer." The expectations were to make waves and develop concepts appropriate to understanding this phenomenon. The goal was nothing less than achieving shared meaning among the wave-related ideas of frequency, amplitude, pitch, energy, wavelength, vibration, interference, and resonance.

Instrument Recipes

How could they provision the room for investigating waves while allowing children to think and express themselves in unanticipated and unintended ways? Could they avoid excessive haste in getting to the objective and reducing understanding to recitation? How might the team teach Physical Science Standard #4 "just so"?

With toys and noisemakers. With splashes and shouts. With strings and straws. With Slinkys and computer simulations and graphical drawings. And a "scientist in residence," courtesy of help from the school's parent community.[8] An old teacher's guide from the 1960s—Elementary Science Study (ESS) *Musical Instrument Recipe Book*[9]—sparked their imagination.

Story Line Essentials

During the planning sessions months earlier the team, in keeping with common practice, had distilled two essential questions to guide instruction about "wave basics" and a story line to organize experiences leading to understanding information transfer. Two essential questions were distilled: "What are waves and how are they created?" "How do wave patterns transfer information?"

The story line told of vibrations making waves happen, then energy transferring from place to place in waveform. "Vibrations" were the central characters in the story, the phenomena of interest. Children tapped tubes, rang bottles, blew whistles, made reeds, plucked strings, and drummed elastic surfaces as they playfully explored ways to invent musical instruments. They tapped on shallow pans of water to set the surface in motion as well as experimented with strings and Slinkys to see waveform patterns. Online simulations of waveforms and their interactions enhanced these experiences.

Although water waves received some attention, sound was the focus of exploration. Children created sounds and learned to describe pitch and volume in association with vibrations. The story line of waves shared examples of animal communication among crickets, frogs, and, of course, elephants and concluded with the idea of coded sounds sending signals and carrying messages. Lessons unfolded over a month in three classes per week, one hour per class, with daily activity recorded in a journal, time permitting. The task of the teachers was to keep attention on the phenomena of interest—vibrations.

The task for the children was to recognize how to represent variation in vibrations as wave drawings initially described in their own vocabulary and, with time and familiarity, in the language of physics. Closely spaced waves indicated rapid vibrations. Drawing tall waves meant the sounds were loud.

WAVE BASICS

"Monday Slinky, Wednesday wave tank, Friday pendulums; all you happy children, we're glad you came to school!" (lyrics to sing to the melody of "Today is Monday" song). Slinkys, strings, and ripples introduced the children to wave basics.

Slinky Time

A Slinky exercise and a wiggly string on a slippery table kicked off week one. For a first act, Slinkys took the stage.[10] Children stood in a circle. To start, Mr. Link asked the children to raise one hand and hold it straight and still, hand open. The preliminary activity was hand-waving: left to right slowly and slightly, completing a cycle in four seconds. Then waving speeded up to two-second cycles and, finally, rapid waving (one, two, three, four/left, right, left, right).

Waving repeated with bigger movements left to right, again slowly then more rapidly. Mr. Lemke introduced the words "frequency" and "amplitude" to describe their brief activity. Obviously, they next performed "the wave" by raising both hands high, then lowering them, following the same action by the person standing at their side, as the wave traveled around and around the circle.

The class separated to stand around their worktables and Mr. Link passed out one-fourth-inch, soft cording cut to the length of the tables. "While the string is laying on the table, hold one end and wiggle it." Wiggles snaked their way down the cord, growing smaller as they reached the open end. Next, pairs of children held and wiggled the string at both ends. "Not too fast!" Wave patterns (squiggle motion) varied.

Everyone next seated themselves on the tile floor, still in a circle. Two volunteers stretched a Slinky across the floor. One student held an end still and the student at the other end, with a quick action of hand and wrist, sent a pulse along the Slinky. The pulse made its way and returned on the other side of the Slinky. This proved interesting and, with teacher prompting, led to finding out what happened when a returning pulse encountered another one sent on its way. Would they cancel each other?

"Things to Do and Demonstrate with a Slinky," especially if a staircase is nearby, could fill volumes.[11] Mr. Link rotated students at the ends of the Slinky and prompted them to make waves. By stretching, they tried making push-pull (compression) waves and by shaking, back-and-forth (transverse) ones. Pulses created by quick shaking and slow shaking seemed to take about the same time to go from one end to the other and back again.

Soon, their coordinated shaking, while sitting, created countable, standing Slinky waves. Four peaks—a fast vibration—went back and forth as they traveled from end to end considerably faster than two peaks—a slow vibration. Getting eight peaks and troughs rapidly oscillating was a challenge—but not impossible.

Mr. Link encouraged the children to express themselves in their own words to describe the motion of the Slinky. He used their language to talk about the Slinky while transitioning to more formal expressions. The context remained the same as he added a problem-solving challenge: "So, the wave 'wiggles back and forth' across the floor. Let's see if we can count the wiggling in ten seconds."

A pattern emerged by recording the number of back-and-forth motions—frequency—in a fixed interval of time. As wave frequency rises, wavelength decreases (the former is the inverse of the latter). The smaller the wavelet, the more the Slinky vibrated. The longer the wave, the slower the frequency—it gets less squealy and more rumbly, like elephant infrasound.

Lessons soon moved to a virtual representation and simulation of a wave on a string coupled to a return to Slinky watching. The intent was to answer the Essential Question and its restatement as, "What are the properties and behaviors of waves?"

The teacher's introduction to Physics Education Technology's online simulation of a "wave on a string" complemented experimenting with both the wiggling string and the Slinky.[12] In preparing to teach, this simulation had proven quite valuable, and the string graphic depicting wave interference patterns mesmerized to the class. Manipulating pulses, whether on the virtual string or the real Slinky, made vibrational patterns in the form of pulses and waves readily apparent and reproducible. The simulation enables manipulating the range of tension and amplitude to generate single pulses or oscillating patterns. Pausing froze the sinuous curves in place for discussion of wave features—amplitude, nodes, wavelength.

Each class leapt at the chance to watch a second (and a teacher's favorite) animation of wave oscillations. The animation presented a graph of two waves (in effect, a virtual oscilloscope screen), one red and one blue, and, if selected, the resulting wave pattern of their interference in bold green. In this simulation the user manipulates wave frequency, speed (analogous to tension in the string simulation), and amplitude (the simulation included choosing whether to start the wave with a node or an antinode, but this distinction was not addressed). Once again, enchantment ruled as the visual display of wave propagation, interference, and standing wave patterns captured their attention.[13]

Note to teachers: wild or chaotic Slinky play guarantees disaster and the virtually impossible task of untangling a kinky Slinky. A playful environment of crafted experiences, used purposefully to introduce language, is not the same as a chaotic, free-choice environment. The important point is that playful exploration enhances opportunities to fashion meaningful questions, whether by teacher or child.

What to Do on Wednesday?

Some tapping on pans of water and dropping small stones to watch waves rippling over an area rather than just marching in a line make good sense. The shallow pan of water clearly reveals the propagation of a wave away from a disturbance. It's also possible to watch what happens when the ripples encounter a barrier—the sides of the pan (or a piece of something solid placed in the pan).

A few dots, the remains of paper subjected to a hole-puncher, sprinkled on the surface behave in an interesting fashion. First, predict what will happen when a ripple reaches each dot. How might the dot respond? Sink? Jump? Float away in the direction the ripple is moving? Slide in the direction the wave came from? Move along the top of the advancing circle of waves? Imagining possibilities primes the focus of attention. Suggesting weird or unlikely motions enriches the investigation. Guessing right is not the point. Looking closely is. (The dots move up and down as the wave passes.)

What to Do on Friday?

Outdoor play is a necessity and recess boasts one of the greatest science teaching apparatuses of all time: the swing, a pendulum of many uses. It goes back and forth, pivoting from its fixed support. For children, the swing offers one of many opportunities to learn physics through play.

Pumping means swinging the legs at the knees, pulling with the hands, and leaning back, then forward. If pumped, the swing swings up from where the hands grip and the knees flex—that's how energy is added. Or a friend can time pushes just right and make the swing go higher and higher while the passenger goes along for the ride. That's resonance, whether labeled so or not. By grade four, children know well the potential to swing higher, then fall back in a rhythmic cycle. They are able to do so with great success and without needing to describe the swing as a resonating system that oscillates as potential and kinetic energy slosh back and forth.

The swing provides a straightforward, yet creative, opportunity to consolidate children's understanding of wave basics. The swing swings forward for a short time, then back for a nearly equal time. With time, and no pumping and no resonating input from a friend, the cycle decays as the pendulum (the swing) comes to rest. The change of position with time is graphable.

The exercise, following exercise on the playground, was to graph the motion of swinging with "time" on the x-axis and "position" on the y-axis. Position means distance in front of or behind the starting position (hanging motionless). Call the forward positions "positive" and the backward positions "negative" to make things a bit more "mathy." There are two directions to plot with respect to time: forward (positive values of "y") and backward (negative values of "y"). Direction reverses midway in the cycle.

For another concrete representation of swinging back and forth as a sinuous curve of peaks and troughs, hang a paper funnel full of salt (or sand) from a support and let it swing back and forth. Then cut a small hole in the bottom of the funnel (the "bob" of the pendulum) so that the sand funnels out onto a piece of paper. As the sand pours out, slowly pull the paper perpendicular

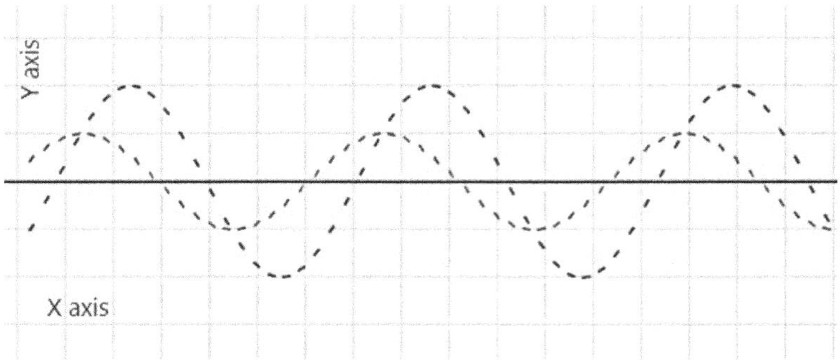

Figure 5.1 Two Waves to Superimpose by Drawing a New Wave Summing Positions Above and Below the X-Axis (Time, in Seconds). Displacement below might stand for a hand-waving in a position left of vertical and above for the hand moved to the right. *Source*: From Tom Henderson, "The Standing Wavemaker," The Physics Classroom. Used with permission.

to the plane of the swinging pendulum. The sand falls as a waveform on the paper, a tangible graph of an abstract idea.[14] Call this contraption a "saltoscilloscope." It advances the story from manipulation to representation. Graphs, by the way, make excellent journal entries, useful as references during discussion phases of instruction.

Speaking of graphs, here's a challenge. Given two interfering waves with similar frequencies, what waveform will they make when superimposed on each other—when they are interfering? What, in effect, is their interference pattern? This is a hard challenge, but it is the basis of the green line representing interference pattern of the blue and red waves in the animation found at the "The Standing Wave Maker" website.[15]

Make a printout of a frozen-frame of animation with the green line set to "off" (see figure 5.1). There should be a few complete cycles of the red and the blue waves on the graph. Along the x-axis pick a point and extend a line perpendicular to this axis until it intersects both waves. Now plot a new point along that vertical line that is the value found by adding the two points of intersection. This value is the sum of two lengths—or the superposition of two lengths, both in the same direction or one in one direction and the other in the opposite direction.

The two points to combine may both be "up" with respect to the x-axis (above zero) or "down" (the negative direction). The task is to find a point, above or below the x-axis, that goes up and up some more, or down and down some more, or up some, then down some, or down some, then up some. The words are easy to stumble over. Do it; it's not hard. Not all, but several Goggins Elementary School fourth graders succeeded.

A CHALLENGING JOURNEY

Learning is a journey where "all of us must cross the line between ignorance and insight many times before we truly understand."[16] Multiple media, multiple phenomena, multiple experiences, multiple discussions reinforced each other to help the children abstract the basic properties of waves demonstrated by the qualities of sound. Teachers focused attention on how vibrations vary.

Pitch and length. Pitch and tension. Count the up and down movements of wave crests on a Slinky in ten seconds, twenty seconds, forty seconds. Shake the Slinky more rapidly, double the number of crests, and repeat the counting. Try to quadruple the original set of peaks. Explore pitch, tone, and volume with tubes, chimes, strings, and whistles.

Perceptions of patterns fuel progress in understanding. Watching animations of wave interference while manipulating variables of frequency and amplitude boosts comprehension. Guided conversation synchs the virtual world to the real one. Insight advances by expressing such relationships qualitatively in linguistic and pictorial form.

On the journey ahead await mathematical formalisms. Awareness that waves stretched to longer lengths go back and forth fewer times in a period of time becomes, in upper grades, the "wave equation" for sound: Speed = Frequency × Wavelength, or $v = f * \lambda$ ("speed" is constant for sound propagating in particular substance, such as room temperature air). In elementary school, children encounter the wave equation perceptually and qualitatively by wiggling strings and watching Slinkies.

Using this knowledge, how might they respond to the challenge of engineering an instrument for transferring information? The *Musical Instrument Recipe Book* gave them an idea: make banjos, with two strings tuned to two pitches. Binary banjos!

NOTES

1. Bill McQuay and Christopher Joyce, "To Decode Elephant Conversation, You Must Feel the Jungle Rumble," Podcast. *National Public Radio Close Listening: Decoding Nature through Sound*, 2015, accessed March 14, 2020, https://www.npr.org/2015/08/20/432616506/to-decode-elephant-conversation-you-must-feel-the-jungle-rumble.

2. Rudyard Kipling, "The Elephant's Child," in *Just So Stories* (Garden City, NY: Doubleday, 2002), 71.

3. Ibid., 76.

4. Jeheskel Shoshani and Pascal Tassy, "Advances in Proboscidean Taxonomy & Classification, Anatomy & Physiology, and Ecology & Behavior," *Quaternary International* 126–128 (2005), 14.

5. Ibid., 13–14.

6. Jeheskel Shoshani and G. H. Marchant, "Hyoid Apparatus: A Little Known Complex of Bones and Its 'Contribution' to Proboscidean Evolution," in *The World of Elephants International Congress* (Rome, 2001), 673.

7. In the learning progress for PS# in the NGSS PS4, secondary students extend their conceptions of waveform transfer of energy and information to include electromagnetic fields.

8. The school's parent association stepped in to fund a "scientist in residence" program, paid for with a budget for supplies and a modest honorarium for hours of planning and several days of co-teaching over six weeks by the guest scientist. The association had previously funded a much-valued artist-in-residence program.

9. Educational Development Center, *Elementary Science Study (ESS) Musical Instrument Recipe Book* (New York: McGraw Hill, 1971).

10. For the story of the serendipitous invention of the Slinky, see Emily Prokop, *The Story Behind: The Extraordinary History Behind Ordinary Objects* (Coral Gables, FL: Mango Publishing Group), Kindle.

11. The Exploratorium, "Slinky in Hand," *Science Snacks*, accessed March 21, 2020, http://www.exploratorium.edu/snacks/.

12. Carl Wieman, founder, and Kathy Perkins, director. PhET (Physics Education Technology project). "Wave on a string" is one of a host of physics simulations for other disciplines created by PhET at the University of Colorado, Boulder. Accessed March 5, 2020, https://phet.colorado.edu/sims/html/wave-on-a-string/latest/wave-on-a-string_en.html.

13. Tom Henderson, "The Standing Wavemaker," *The Physics Classroom*, accessed March 23, 2020, https://www.physicsclassroom.com/Physics-Interactives/Waves-and-Sound/Standing-Wave-Patterns/Standing-Wave-Patterns-Interactive.

14. Educational Development Center, "Salt Pendulums," in *Elementary Science Study (ESS) Pendulums: Investigating Physics with Swinging Objects* (New York: McGraw Hill, 1976), 20–21. Recording the salt patterns from salt (or sand) pendulums can be modified by pulling paper to display the sinuous waveform pattern.

15. Henderson, "The Standing Wave Maker."

16. David Hawkins, *The Informed Vision: Essays on Learning and Human Nature* (New York: Algora Publishing, 2002), 70.

Chapter 6

Binary Banjos

"Squawk, squawk" said the little red hen. Not a real hen, of course, but a red plastic cup with a string hung inside, a small piece of wet sponge tied to the end of string, and a bit of yellow paper cut into a triangle, glued to the cup—a chicken's beak! Tug on the string with the sponge and voilà, squawking happens. Some might say, "Sounds more like a turkey to me."[1]

The squawking-hen system is perhaps the world's simplest banjo design. It's a string and a drumhead. Tugging on the string makes it vibrate—a bit like rubbing chalk on a blackboard (old enough readers will recognize this reference). The cup vibrates, too, but seems able to sing only a single note. The standards challenge students to engineer a device demonstrating an understanding of how waves transmit energy and carry information. The little red hen (red if the cup is a red one) clues how to do so. But before banjo-making begins, there's more to learn about pitch.

PITCHING SOUND

With careful technique, someone blowing across the top of an empty soda bottle (soda, not beer bottles, for school) can make quite a whistling sound—high and low pitch, same bottle. What happens to pitch when water is added? How does the sound change when blowing harder or softer—when adding more or less energy?

Plucking and tapping also make things vibrate. Whether testing strings, tubes, chimes, or reeds, the curious fourth-grade investigator can learn to control pitch independent of loudness.

Whistling Air

Flutes, recorders, and organ pipes are similar to whistles. In each, air moves briskly across an opening with an obstruction (the edge of the bottle, for example), setting the air in a chamber vibrating.[2] How do the sounds of vibrating objects vary? What factors change pitch? Tongue depressor sticks clamped to a table, cardboard tubes of different lengths to blow across, and much more invite playful exploration of pitch and vibrations.

Focus on listening as water pours from a bottle into a cup. Listen closely to sounds coming from the bottle. Does the pitch rise or fall? Pour the water back into the bottle and keep listening. Repeat a few times. Pitch falls when pouring water out of a bottle. Pitch rises when pouring the water back in. Amazing! Pitch depends upon the air volume in the bottle: more air, lower pitch.[3]

Plucking Strings

Classroom tables make fine bass instruments when properly strung. Take one rectangular table; place a meter stick across it (or perhaps a wooden dowel or a one-inch by two-inch wooden slat). Circle the table lengthwise with a length of strong cord crossing the meter stick (the bridge) and tie in a loop beneath the table.[4] Tighten the cord like a tourniquet by twisting a ruler placed through the loop to produce higher and higher pitched sounds. Use a clip to fasten the ruler so that it does not unwind. If feeling ambitious, wrap multiple loops around the table and tighten them differently. "Be careful that the meter stick doesn't fall over," Mr. Link cautions, "It's the bridge and needs to stay in place."

Try placing bridges at each end and slant them across the table. Move each bridge away from the edges of the table, thus dividing each string into segments of different lengths. The table will begin to resemble a traditional folk instrument, the hammered dulcimer. Hammered dulcimers have two separate sets of strings: one over the base bridge and the other over the treble bridge; teachers enamored of music theory as well as fourth-grade musical prodigies may be aware that across the treble bridge the strings are tuned to a perfect fifth.

Now pluck. "Thump, thump" goes the classroom table. How might a capo (abbreviation of the Italian "capodastro," a bar pressed against a fret) enrich this experience? Do add frets if so desired. The addition of a thin dowel or two beneath the strings will do. The point is to have children design and engineer some plucky instruments of their own. Children may have other ideas for making musical furniture. Of course, the old-fashioned way makes a fine base: broom handle, cord, and inverted, galvanized tin laundry tub, as featured in jug bands. An empty, rigid plastic, five-gallon paint container will do if laundry tubs are scarce. Sans paint.

Keep in mind that simply stretching a couple of rubber bands over small cardboard boxes produces suitably annoying sounds. Try a matchbox. Vary the bands.[5] Use these to accompany the squawking chicken diva in a live, fourth-grade opera. No rehearsal necessary; just perform (supposedly when other nearby classes are not engaged in sustained silent reading).

Tapping Chimes

Tap on metal tumblers filled with water to various levels. A simple challenge focuses attention. Do the chimes sound the same when tapping above and below the waterline? (Glass works, but is hazardous; besides, inexpensive metal tumblers have many elementary science uses.)

Speaking of chimes, the most intriguing ones resonate with tones and combine them with overtones. Perhaps the most impressive is the sound of overtones made by "spoon-chimes."

Attach a strong thread or length of kite string to a spoon. Or to several differently sized spoons for added enjoyment. Strike the spoon with another utensil and listen while suspending it from a finger, thread wrapped not too tightly around the finger. Gently place the finger against the opening of the ear canal—not inside, just enough to touch. Be careful (and do not share ears and fingers for hygienic reasons). Listen again to the spoon chiming when struck. Or to any other utensil hanging by a thread. Does a rolling pin work to make a chime? A plastic utensil?

Now it's time for the performance by the celebrated Oven-tones. Obtain an oven rack. Tie string or thread to two corners, wrap the threads around two fingers (one on each hand). Lean forward, fingers in ears. Now ask an assistant to strum the rungs of the rack slowly and lightly. Where does the sound seem to come from? Can the sound be put into words, or is it just too amazing? Might there be other sound experiments to be had hanging by a thread?

Of course, garden chimes may be displayed. Copies might be engineered from polyvinyl chloride tubing or inexpensive electrical conduit cut to different lengths to demonstrate grasp of the relationship of wavelength (tube length) to frequency (pitch). This relationship can be pursued at virtually any level of sophistication because pipe instruments can be finely tuned. Tubes (pipes) are chimes when struck. In organs and flutes, air vibrates within tubes.

Tuning Tubes

There is a simple way to construct an infinitely tunable, single pipe to make a flute or slide whistle. Obtain a length of flexible, clear plastic tubing, one-half inch in diameter and a length of two or three feet. Mouths will touch one end, so prepare for proper disinfecting after each use by a classroom musician. Best is simply to have the teacher demonstrate the instrument.

Hold the tube with one end closed (thumb, finger, or cork will do). Trickle in some water to fill the tube more or less half full. Hold the tube like a flute, touching the bottom lip to the edge of the open end. Lower the hand holding the closed end. Blow across the open end to make a whistle. What will happen to the water level as the flutist lifts the closed end, forming a large U-shape? Well, the water level rises, of course. Keep blowing while raising and lowering the closed end. The pitch when the air column gets quite small becomes quite high.

For humorous effect, lift the closed end until water reaches the top of the open end. The high-pitched squeal ends in a spray of water. Anyone nearby will get sprayed. It's advisable to turn sideways to the audience and maintain a safe distance.

Plastic tubing and funnels make for interesting play, communication, and instrument design. The fourth graders at Goggins Elementary School indeed had fun speaking to one another from the playground to the library by means of the "funnel-phone," a variation on the classic tin-can-telephone (vibrating air in the tube carries sound similarly to the vibrating string).[6] It stretched thirty feet, a large plastic funnel inserted in each end of the three-fourth-inch tubing.

Voice traveled well; caution: too well can be a problem. No shouting! How soft a whisper can be heard? The funnel is a "reverse megaphone" at the speaking end. (Again, sanitary cleaning is necessary when switching from one broadcaster to the next.) Split the main line with a Y or T junction. What happens to volume at the listening end?

Short sections of tubing with a small funnel as a mouthpiece and a larger one as a bell make for good substitutes for trumpets in the plastic section of the class orchestra.

Journal entry challenge: draw a diagram of what happens to air in the tube. Draw a diagram to explain how we hear a voice from across the room (not through a funnel phone). Use the diagrams for discussion; do not worry about accuracy.

What's moving? Where are the spoken words being formed from one moment to the next? What's the air doing there? If someone speaks from across the playground while a crosswind blows, does the sound of the speaker's voice move sideways with the wind? Maybe revisiting and revising explanatory drawings will help sort out a plausible answer. Where's the sound at time one, time two, time three, and so on, until heard? Being correct does not matter less than careful thinking and explicit expression.

Up next: a silly French horn, fashioned from plastic tubing.

Buzzing Reeds

Recently, conservation efforts have placed plastic straws on the endangered implements list. For playful experimentation with sound and vibration, from a musical instrument point of view, they remain a necessity.[7] With two cuts, taper the end of a plastic drinking straw. It's a good idea to make another cut to remove just the sharp tips of the tapers. With trial and error, and a bit of practice, pinching the tapered ends close together with the lips while blowing through the straw makes a pleasantly irritating, buzzing sound. Lips feel the vibration; it's a reed! Cut shorter and shorter, the pitch rises. Notches cut in a full-length straw make playing different notes possible.

Now cut a length long enough to insert snugly into the end of several inches of plastic tubing. Notice how the frequency slows down dramatically. Instruments made with even longer lengths drop the pitch more and more. Add a funnel to bell the horn, coil a rather long tube, and create a truly low and slow, repeating note. The room comes alive with the sound of flatulence. Expect this outcome to produce animated responses from children.

THE BANJO'S STORY

The word "banjo" immediately raises an interesting philosophical question: "What makes a banjo a banjo and not a guitar?" A banjo is a tambourine without symbols, plus some guitar parts. It starts as a drum—a stretched parchment connected by a bridge to plucked and vibrating strings. Similar to a guitar, banjos have long necks to extend the strings that cross a bridge and frets to change pitch (and key), and, of course, tuning pins to adjust string tension. Resonating chambers to amplify sound are essential components of both instruments.

African Drumming

Africans, torn from consummate drumming cultures, created banjos in America from gourds.[8] In Africa, drums had for millennia broadcast complex messages from one society to another through variations in rhythm and melody that mimicked patterns of speech, a technology not understood by Europeans until John F. Carrington, an English missionary who made Africa his permanent home, published *The Talking Drums of Africa* in 1949.

After a day of walking through the Bambole forest along the Upper Congo River, young Carrington arrived in Yaongama and was astonished that the townspeople had gathered waiting to greet him. How had they anticipated his arrival so precisely? Drumming. Through the years he learned how skilled drummers "talked" in a language of beats and tones to share "poetry, prayers, and jokes."[9]

Why had the language of drums eluded non-Africans? Carrington devoted his life to the study of African languages and dialects to find out. He understood the dependence of most African languages on tonal qualities: patterns of rising and falling pitch. Most Indo-European languages do not invoke pitch to establish meaning, the exception being signaling a question. Tonal patterning is not unique to African languages; contouring pitch permeates Mandarin, for example. Carrington eventually learned not to say, "He boiled his mother-in-law" in place of "He watched the riverbank," despite the phrasing of both statements with the same phonetic syllables, "Alambaka boili."[10]

At Goggins Elementary, this dimension was a fascinating footnote, set aside as a possible topic to investigate at another time or in another cycle through communication and wave patterns. At another school in the same city, however, African drumming was the starting point for an integrated study of culture, history, music, and science. In place of a scientist-in-residence the school hosted a musician-in-residence (and a long-term visit from a skilled drummer from Nigeria) who led a curriculum entitled "Under the Baobab Tree." Clearly, in the banjo's journey from African forests to Appalachian hills (the banjo has become an American folk instrument) there are many dimensions to learning science beyond standards.

Recipe for a Binary Banjo

Even though African drumming failed to find a place in the science lessons this go-around, thunder and elephant infrasound serendipitously did make a story line appearance. If animals can decode sound patterns signaling the distant whereabouts of herd relatives, then certainly fourth graders can engineer communication devices with similar capabilities. The trick is to control vibrational patterns—and then decode the pattern. Listening to a musical instrument is a good example (as is listening to someone talking, of course), but instrumental music, without lyrics, communicates mood and emotion—with harmonies, overtones, and rhythm.

The Elementary Science Study's *Musical Instrument Recipe Book* once again proved to be the key resource. There the teachers found inspiration: ideas for making a simple banjo (more complex and true banjo designs as crafts projects, using a cookie tin as a resonator, for example, are out there in the land of Google).

Sturdy, large plastic jugs served as resonating chambers (not drumheads, but twangy nonetheless). Sawing converted paint stirring sticks into bridges (a teacher worked after school with a glue gun to put them in place; wooden dowels might have worked better). Sturdy lengths of one-inch by two-inch slats of wood formed the necks (cheap yardsticks are too flimsy). Screw eyelets attached monofilament to the ends of the slat; twisting them adjusted

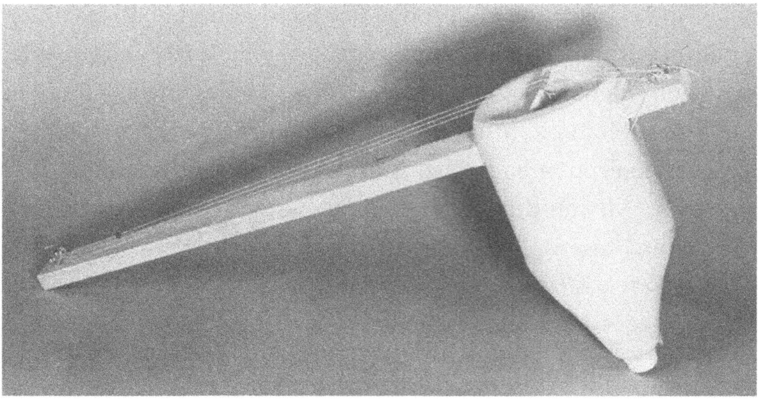

Figure 6.1 The Original 1968 "Bleach-Bottle Banjo." *Source*: From *ESS Musical Instrument Recipe Book*, copyright © 1968/1971 by Education Development Center, Inc. (EDC). Used with permission and with all other rights reserved.

tension. Trial and error worked wonders, but the banjos were fretless (see figure 6.1).

Plucky Codes

Each child made a banjo. They all worked—and parent volunteer help at school (not home) had been a big factor. Now came the challenge of messaging with pitch and creating a system for decoding messages. Two strings, adjusted to two different tensions, made for two pitches. On-off, up-down, yes-no, one-zero: these are binary (two-value) systems. Coding pitch with binary numbers offers an attractive way for talking in banjo. First, of course, the teachers had to teach their classes (as well as review for themselves) how to count with binary numbers.

In two classes 1s were assigned to high pitches and 0s to low pitches. Binary coding the alphabet assigned the value 00001 to the letter "A." "E," the fifth letter of the alphabet, became 00101.

The third class went in its own direction and assigned letters using Morse code. High pitches became dots; low pitches, dashes.

What to Say and How to Say It

A single stretched string segmented in two unequal lengths can produce two pitches, of course—as can one string with a fret. However, alternately plucking two strings of different tension, hence pitch, does feel more musical (and it's quite efficient); "0" for the low pitch; "1" for the high pitch—or vice versa: either is fine, so long as sender and receiver are in agreement.

Coding learned, writing short personal messages ensued. A lucky student was chosen to go first. A hush fell over the room (an unusual spontaneous event among fourth graders), then five twangs, a mix of high and low pitches, broke the silence. "01100," low, high, high, low, low. The banjo player paused as the class decoded using the key projected on the screen. "M" wrote everyone (00001 stood for "A" in this class). "Y" followed, and, at last, everyone could read, "MY NAME IS ANNA."

Children had found their way from elephant stories to recipes for musical instruments garnished in binary code. Playfulness and pleasure dominated their school science journey.

NOTES

1. Science Bob Flugfelder, "Chicken in a Cup," accessed March 13, 2020, https://sciencebob.com/chicken-in-a-cup/.

2. Jearl Walker, *The Flying Circus of Physics with Answers* (New York: John Wiley & Sons, 1977), 16, 230.

3. Ibid., 14, 229.

4. Victor E. Schmidt and Verne N. Rockcastle, *Teaching Science with Everyday Things*, Second Edition (New York: McGraw Hill, 1982), 134.

5. Ibid., 5, 226. Rubber bands do not respond the same as guitar strings. Tightening a guitar string raises pitch. Stretching a rubber band some, then some more, does not change pitch very much at all. If anything, stretching may lower the pitch! Frequency depends on a string's tension (tightness), density, and length. Tightening tunes a string by increasing tension. It does not change length or density (at least not significantly). Stretching a rubber band changes all three properties at once. As a result, frequency changes very little—probably unnoticeably.

6. Deceptively Educational, "Fun with a Funnel Phone," accessed March 14, 2020, https://deceptivelyeducational.blogspot.com/2013/04/fun-with-funnel-phone.html.

7. Steven Dufresne, "How to Make Music Instruments Using Straws—Trombone and Trumpet," accessed March 14, 2020, https://rimstar.org/science_electronics_projects/make_music_instruments_with_straws_trombone_trumpet.htm.

8. Stephanie Kingsley, "Chronicling America's 'African Instrument': Laurent Dubois on the Cultural History of the Banjo," *Perspectives on History Perspectives*

on History (The American Historical Society, June 19, 2017), accessed May 17, 2020, https://www.historians.org/publications-and-directories/perspectives-on-history/summer-2017/chronicling-americas-african-instrument-laurent-dubois-on-the-cultural-history-of-the-banjo.

9. James Gleick, *The Information: A History, A Theory, A Flood* (New York: Random House, 2011), 22.

10. Ibid., 23.

Chapter 7

Harmful Haste

One Friday, at the end of a lesson on how to separate components of mixtures and solutions, a class of third graders stirred salt into cups of water, poured some onto saucers, and left the saucers on the windowsill to dry over the weekend. On Monday, the children looked closely at the saucers. They had the advantage of observing them with magnifying glasses. The teacher asked, "What happened to your saucer of salty water left on the windowsill over the weekend?"

Sam timidly raised her hand. When called upon, she answered, "The water evaporated and left a solid solution."

SOLID THINKING ABOUT SOLUTIONS

Pleased by the use of the term "evaporated," the teacher proceeded to write Sam's words on the board, but stopped at the end of the phrase, "The water evaporated." So far, so good, thought the teacher. What Sam said that followed—left a solid solution—the teacher had neither anticipated nor accepted as correct. Talk about a "solid solution," the second part of Sam's answer, did not conform to the teacher's (and to the teacher's guide's) expectation. For Sam, "solid solution" was a creative way to share what she saw. For the teacher, truncating Sam's language was a way of offering mild correction and gave others a chance to contribute.

Thanks to others in the class, the teacher heard what she was listening for and added to the statement on the board ". . . and only salt crystals remain."

Hmmm. What had Sam observed? Under magnification, many of the little squares (cubes, actually) of salt appeared to hold droplets of water in their centers. Fluid inclusions in crystals are hardly rare in nature and they were common in the crystals grown over the weekend by the third graders in their saucers. The moment passed for accepting Sam's language for describing the droplets within the crystals. Sam had answered neither incorrectly nor nonsensically; her words simply did not agree with the scripted "right answer."[1]

Sam had cobbled together words to describe a novel phenomenon and accomplished a very respectable act of linguistic invention, "solid solution." She spoke, unwittingly of course, in the jargon of chemistry subsuming a precise geologic term. Labeled "feldspar minerals," solid solutions of oxygen, silicon, aluminum in varying combinations with sodium, calcium, and potassium constitute 50 percent of the earth's crust. Solid solutions are real and hardly rare, just difficult to imagine for people encumbered by right-answer science teaching.

Of course, few elementary school teachers have a background in mineralogy or an in-depth grasp of chemical equilibria. The notion of "solid solution" was more sophisticated than the information presented in the teacher's guide. The teacher had, in executing the curriculum adopted by the school district, skillfully organized a comprehensive sequence of lessons for teaching the concept of a solution.[2]

Correcting the child's language, born in part from insecure understanding, kept the class moving in the expected direction—but with haste. Whatever the teacher's level of understanding of science, humility and confidence, the parents of patience, work well together. A classroom embrace of playfulness welcomes children's original utterances and places trust in patience.

At the back to school night preceding the study of mixtures and solutions, parents learned that children in the third grade would be learning "the subsystems and variable concept." Few in the audience had a clue as the meaning of this utterance. One did express enthusiasm for the idea of children becoming familiar with proper scientific vocabulary, even without understanding, in order to be better prepared to learn science in upper grades.

Standards become destructive of children's interest when reduced to watering down upper-grade objectives for learning science. Exposure to "proper" vocabulary, and care to hit and cover topics in early grades to be better prepared in upper grades, poisons elementary science.

A child's expressive language belongs in a science class. Children play with ideas by creating novel amalgams of familiar and unfamiliar language. They try out expressions to see how they fit puzzling circumstances. Linguistic creativity is an excellent outcome, not a wrong answer. That's a lesson for teaching science resting upon a heritage of over 250 years.

THE BROKEN STICK

The year is 1762, and a French philosopher imagines himself tutoring a youthful pupil named Èmile. The day's science lesson begins by directing Èmile's attention to a clear glass of water. Into the water the philosopher drops a stick.

The philosopher comments, "The first time a child sees a stick dropped halfway in water, he sees a broken stick."[3] Is the stick actually broken? Of course not. Being a philosopher, Èmile's tutor rejects the temptation to explain in haste why the stick looks broken. He seeks, instead, to uncover a universal truth in this particular episode of instruction.

The philosopher speaks to his reader and asks, "To know if it is broken, as it appears to be, how many things must we do before drawing it from the water?" Drawing it from the water would reveal that, in fact, the stick remains unbroken. Maybe Èmile should spend some time looking closely from many angles. How does the stick's appearance change? The philosopher expresses the educational principle appropriate to the situation: "I would prefer that Èmile never know dioptrics [the optics of a two-lens telescope] if he cannot learn it around this stick."[4]

Telescopes and microscopes make distant objects appear close and tiny ones bigger. The "dioptrics" of both functions to place the observer closer and closer to the object of interest. They magnify. Look again at the twig or pencil in the cup of water. Broken, yes, but also magnified below the surface.

Stick, twig, or pencil, even a sideways glance makes the object appear broken at the surface of the water, the top portion separated from the bottom. That's a forever truth, a puzzling situation that invites interest. It looks broken; how can that be? Maybe it's magic and after a moment of surprise it's time to move on. But this is a science lesson and a time for play.

FUN HOUSE IN A CUP OF WATER

Replicating Èmile's conundrum in the 2020s is quite easy. Use a clear plastic cup filled nearly to the top with water; a six-ounce size will do nicely to begin. Place a pencil in the cup leaning to the side. Now look at the pencil. Then look at the cup from directly above. The appearance of the pencil changes: broken at the surface of the water and magnified beneath. The shape of the cup is a circle and thus, filled with water, it becomes a lens. If the cup's diameter decreases with depth, then the curvature of the lens increases.

Move the pencil and watch its appearance change. In no time students will see that it appears to bend like rubber into a sinuous shape. The image of the pencil reflects from the water's surface when viewed from the proper angle because the simple cup of water is both lens and mirror. Distortions, reflections, and "breakage," oh my!

Go ahead and draw the pencil out and put it back in the water many times. Think of how many different ways there are to look at the pencil using the cup of water. The pencil might be *in* the cup, *behind* the cup, *above* the cup . . . lots of prepositions work to suggest how to manipulate the relative position of the pencil and the cup of water. If there are thirty students in the class, they can probably find thirty different ways to position the pencil. If they become stuck at a dozen, add another pencil or a second cup of water.

At table level, look through a first cup of water lined up with a second one. Place the pencil in the second cup. Move it around. Try cups of different sizes. Add real sticks that branch into twigs!

Return to the original problem: hold the pencil vertically in the cup of water. Move it slowly from front to back, then side to side, and finally around in circles. The effects are mesmerizing. Lenses and mirrors can do crazy things to the path of light. Vision detects only light emanating from its last point of departure leading directly to the eye, not its true starting point, a phenomenon magicians know well how to exploit. From this final pathway, the brain extrapolates this pathway back to a supposed point of origin—and not necessarily the original source, the edge of a pencil, for example. Things appear as and where they are not. Much about light and optics can, indeed, be learned from a pencil, finger, or stick "broken" by a cup of water.

The formal concept, by the way, is "refract." The twig appears broken because light refracts (changes direction or "bends") as it crosses the boundary between one clear material and another, a phenomenon to a first approximation explained as due to the different speed of light passing through each material. Right away reading a brief attempt at an explanation will likely be off-putting to many readers—especially those without confidence in their knowledge of optics.

ROUSSEAU'S ADVICE

Playful exploration in response to the challenge to make the pencil look broken or funny or strange or different keeps things moving along. What would Èmile's philosopher have us do?

> First, we walk around the stick, and we see that the break turns as we do. . . . We look from straight above . . . the stick is no longer curved. . . . We stir the water's surface. We see the stick fold up in many pieces, move in zigzags. . . . We let the water flow out, and we see the stick straighten out little by little as the water goes down.[5]

Remarkably, the stick zigs and zags—an astonishing phenomenon and a moment worthy of savoring, pondering, and sharing. Hastening to explain the puzzle interferes with the value of such moments. The contrast—between explanatory squelch and exploratory fascination—brings to prominence another insight from the philosopher: never hurry to satisfy curiosity! His most significant admonition is to avoid haste; he advises the teacher to encourage the pupil to stay attentive to phenomena and:

> Put the question within reach. . . . If you ever substitute authority for reason, [your pupil] will no longer reason . . . [and will become] nothing more than the plaything of others' opinion.[6]

Jean-Jacques Rousseau's *Émile or On Education* stands as a milepost in the heritage of science teaching. *Émile* is not the last word, the only word, or even the first word guiding the field. Rousseau's voice, however, must not be overlooked. It anchors the commitment to playful exploration as an essential dimension of learning science. "Present interest—that is the great mover, the only one which leads surely and far."[7] The two—present interest and playful exploration—are inseparable buddies.

Rousseau expressed contempt for instruction that readily rewards children who excel at rote learning—those who earn praise for becoming quick to offer short, right answers. He warned that "apparent facility at learning" may turn into "the cause of children's ruin." The rote learner's "brain, smooth and polished," he continued, "returns, like a mirror, the objects presented to it. But nothing remains; nothing penetrates. The child retains the words; the ideas are reflected off."[8]

Digital platforms, no different from in-person classrooms, are fully capable of treating the mind as mirror, the child as empty vessel. Most any technology may function perfectly well as an instrument of rote learning if depended upon to hasten understanding unmoored from interest, experience, and pleasure.

To believe, however, that spontaneous involvement in exploration, and from this involvement, an increase in knowledge can be sustained by children's nature alone is a romantic fallacy:

> Good teaching is a many-sided professional activity which indeed often leaves children to their own resources—partly to discover what those resources are. On the other hand, good teachers sometimes intervene actively to redirect the work and sometimes join in an investigation as partners in the enterprise. The overriding purpose is that children grow in knowledge and organization, discipline, and independence of mind and character.[9]

Benign permissiveness—something Rousseau in fact advocated—coupled to notions that children develop best when left alone in an enriched environment discounts the value of active intervention by teachers. In his essay, "Nature Closely Observed," mathematician, philosopher, and educator David Hawkins acknowledged an essential principle tethered to Rousseau's insight, but not anchored to them: "Children should *often* be instructed, but they should *seldom* be instructed before they are prepared to engage, to criticize, and test at least some small part of what they are being taught" (Emphasis in original).[10] Playful exploration endures because play enhances perception, promotes engagement, and delivers enjoyment. A foundation of playfulness, often revisited, supports the disciplined pursuit of interest and the growth of confidence in oneself as learner.

THE CHILD IS PARENT TO THE SCIENTIST

Stories of children becoming scientists underscore that the meaning of an experience cannot be contained by the moment in which it happens. Ideally, adults hold on to their playful, inner child. Childhood curiosity follows many scientists into adulthood, and Charles Darwin's life is an excellent example.

As a young child, Charles Darwin profited from his educational liberty to become "an inveterate collector and hoarder—of shells, postal franks, birds' eggs, and minerals."[11] Following the death of their mother, Darwin's sisters, Caroline and Susan, turned to the teachings of Johann Heinrich Pestalozzi, a follower of Rousseau, for guidance in raising their younger siblings. Pestalozzi counseled, "The child must be led to see for himself that which he is to learn, and not to take it upon the mere authority of the teacher."[12] Later in life, Darwin's disciplined learning transformed his playfulness into inquisitive habits of mind.

A young child boards an aircraft for the first time. The morning is drizzly and the sky dark grey. The plane takes off and climbs through the clouds. After a few moments, the child looks out the window and exclaims, "The clouds are white on top!" The child's parent shares in this delight and asks, "Are they white just on top, or part way down, too?" The child gazes at the clouds for an extended period of time, wondering private thoughts. Years later the child devotes a career to the study of the physics of the upper atmosphere. Did a first airplane flight seed this interest?

MR. POTATO HEAD

Mr. Potato Head, a well-known supporting actor from *Toy Story*, has his own exhibit in a corner at the Strong Museum of Play in Rochester, New York.[13] In the beginning, he was no more than a few pieces of plastic (ears, lips, mouth, eyes, and mustache) to poke into a spud—or maybe a turnip if being creative. Now he's a plastic vessel with a snap-open compartment in his derriere for storing facial features. Television advertising made his designer very wealthy in a short period of time.

Important to remember is not the evolution of Mr. Potato Head, but the message about play at the Strong Museum. Children enter the world ready and driven to play. Children anticipate play with pleasure, especially when it holds surprises—and laughs (think Mr. Potato Head). As stressed by the exhibits of the Strong Museum, a child's sense of strength thrives on whimsical play. Play provides a firm footing for serious understanding, and confident understanding leads to poise. In science, poise turns joyful insights into confident cleverness, the close sibling of playful exploration. Gallery after gallery at the Strong Museum, including a recreation of Sesame Street, reinforces this message.

Playfulness, whimsy, and freedom from fear are basic to learning, and leading works better than dragging. At all levels of "work guided by thoughtful and responsive teachers, there is a spirit of play."[14] Progress depends on imagination and haste does harm; talking and tinkering need ample time.

Children take pleasure in play. It places them in mostly predictable situations that welcome surprise. The actions of play render the unfamiliar, familiar and, at the same time, make possible turning the familiar into something novel. Like a scientist confronted with anomalies that defy established ways of interpreting nature, the child's explorations are fluid, and playful exploration is the proper antidote to insecure knowledge.

Cooperation and collaboration are essential to many forms of play as well as to the conduct of science. In play, as in learning, tolerance for frustration must match the nature of the challenge. Friends (like colleagues) help. Good teachers understand the value of playfulness. They listen and watch children with care, allowing them latitude to think and imagine free of censure. Good teachers refrain from correcting the utterances of children engaged in interpreting their own experiences. They recognize that "play is training for the unexpected," a trait wired into the biology of humans and other mammals.[15]

NOTES

1. Charles R. Ault Jr., Michael R. Cohen, Larry Flick, Christine Kuehn, Joseph Stepans, and Gilbert Twiest, "Interpreting Children's Misconceptions in Science: Responding with Dismay, Alarm, or Delight to Theory, Romance, or Error," panel presentation at the National Science Teachers Association (NSTA) national convention, Seattle, Washington, April 6–9, 1989.

2. Robert Karplus, Herbert D. Thier, and the Regents of the University of California, Berkeley, "Subsystems and Variables," Science Curriculum Improvement Study, Grade 3, Physical Science Sequence (Chicago, IL: Rand McNally & Company, 1972). The original SCIS lessons for Mixtures and Solutions included now-banned activities with Freon. Mixtures and Solutions became a module in Science Activities for the Visually Impaired/Science Enrichment for Learners with Physical Handicaps (SAVI/SELPH, 1983), accessed May 15, 2020, https://files.eric.ed.gov/fulltext/

ED344755.pdf. Subsequently, the unit was incorporated into the Full Option Science System (FOSS, 1993). Elves at the University of California Berkeley's Lawrence Hall of Science developed and promoted SAVI/SELPH and FOSS.

3. Jean-Jacques Rousseau, *Émile or On Education*, trans. Allan Bloom (New York: Basic Books, 1979), 204. Originally published in 1762.

4. Ibid., 206. Dioptrics is the study of refraction using a lens. Optical systems that combine refraction and reflection are referred to as "catadioptrics" ("catoptrics" is the study of curved mirrors). The cup of water is a catadioptric system or a dioptric one, depending upon how you look at it. A telescope may use only refracting lenses or a combination of curved mirrors and a refracting lens.

5. Rousseau, *Émile*, 205.

6. Ibid., 168.

7. Ibid., 117.

8. Ibid., 106.

9. David Hawkins, "Statement on Environmental Education before Senate Education Subcommittee," University of Colorado Boulder Libraries, Special Collections & Archives, David Hawkins papers (May 19, 1970): identifier: COU:758:002:002:0014, accessed May 15, 2020, https://archives.colorado.edu/repositories/2/archival_objects/18648.

10. David Hawkins, "Nature Closely Observed," *Daedalus* 112, no. 2 (Spring, 1983): 74, accessed April 16, 2020, www.jstor.org/stable/20024854.

11. Adrian Desmond and James Moore, *Darwin: The Life of a Tormented Evolutionist* (New York: W. W. Norton & Company, 1991), 13.

12. Johann Heinrich Pestalozzi, in Randall Keynes, *Darwin, His Daughter, and Human Evolution* (New York: Riverhead Books/Penguin Putnam, 2002): 111, quoted and cited from E. Woodall, "Charles Darwin," *Transactions of the Shropshire Archaeological and Natural History Society*, 8 (1884): 14.

13. The Strong Museum of Play also has a Slinky exhibit with its own set of stairs. In early 2021, the Hasbro company renamed their toy spud Potato Head—a gender-neutral moniker.

14. David Hawkins, "Nature Closely Observed," 78.

15. Marek Spinka, Ruth C. Newberry, and Marc Bekoff, "Mammalian Play: Training for the Unexpected," *The Quarterly Review of Biology* 76, no. 2 (2001): 141–168.

Chapter 8

Serious Whimsy

"Fish breathe in the water. They breathe the same way we do, but they breathe out bubbles instead of words," first grader Henry explained one day to his teacher. In just two sentences Henry expressed a remarkable set of ideas. Few experiences are more delightful than listening closely, and without judgment, to children talking about the natural world. Maybe Henry has more to say—or ask—about how speaking and breathing are similar.

Disarming thoughts and questions cascade from the mouths of children, especially when they feel respected, accepted, and confident. Children seem to hoard "why" questions. Why do cats have whiskers? Is it light or dark inside your eyes? Do clouds float? Why is ice slippery? If you get close to the Sun, will gravity pull you in or will rising heat push you away? Why did dinosaurs get so big?

Within every child there is an Elephant's Child, Rudyard Kipling's character from the *Just So Stories* (1902) he told his daughter at bedtime. The Elephant's Child "was full of 'satiable curtiosity, and that means he asked ever so many questions. . . . He asked questions about everything that he saw, or heard, or felt, or smelt, or touched."[1] Not only young elephants but also children are quite capable of asking an overwhelming range of questions. Kipling was a master at answering them with seriously whimsical stories.

MESSING ABOUT

Trying to corral ever so many questions would seem an insurmountable task. Yet children's questions, whether spontaneous or prodded, deserve a prominent place in science teaching. The trick is to bring to attention the right question at the right time—to manage encounters, interactions, and dialogues among children, their teachers, and phenomena while maintaining a playful match between seriousness and whimsy.[2]

In 1965, mathematician David Hawkins, inspired by his preschool teacher wife, Frances Hawkins, penned an immortal class of science education: *Messing About in Science: Phases O△▢*.[3] The phases are messing about, multiple pathways, and lecture/discussion. The shapes symbolize that the phases have no rigid order; any one can lead to any other and back again as many times as necessary—necessary in terms of making learning the search for an appropriate and compelling answer to a meaningful and challenging question. A science teacher's motto ought to be, "Teach the question!" and start with its creation.

Messing about alludes to the delightful and whimsical children's book, *Wind in the Willows*. In the story Ratty (a vole) and Mole drift leisurely along the Thames, appreciating that "simply messing about in boats," with no particular destination in mind, is a most commendable way to pass time together. Teaching science beyond standards stands to profit from the wisdom of a vole and a mole just messing about in a boat.

Phase *O*—"messing about"—too often gets shortchanged. During this phase, the teacher intently observes children's open-ended, playful encounter with objects and phenomena of interest. If done successfully, children (and the inner child of any adult) sharpen their perceptions, make puzzling discoveries, and start to frame questions of their own to share with classmates.

Messing about is not the same as free-choice learning. It can begin with a challenge crafted by the teacher, or simply by drawing attention to an anomalous event. During the heyday of the "open-classroom" when romantic excess of belief in children's innate curiosity and a fondness for "discovery learning" were in vogue, many overlooked the three-part nature of Hawkins's program. He did stress a "lecture phase" (▢)—a much-needed component, best pursued while crafting interest by means of the *O* and *△* phases. By the end of the twentieth century, the primacy of literacy aims in elementary teaching built upon Hawkins's phases to become "science talks"[4] and "science symposia."[5]

Messing About with Pendulums

Pendulum motion, an *Elementary Science Study* staple, was one of Hawkins's favorite objects for messing about.[6] Messing about—noticing, wondering—with pendulums needs ample time prior to conducting an experiment in the classical, single-manipulated-variable mold. Some pendulums are inverted (think metronome or a water tower in an earthquake). Some are coupled (two separate, hanging pendulums with a connector between their strings).[7] And some are "bobless"—the mass is spread out over the length (think of a baseball bat). Researching the mathematical patterns that describe pendulums connected with springs still occupies the frontier of science.

Schools, unfortunately, hastily push students to isolate a single variable that determines the period of a pendulum, whether they are convinced it's a true relationship or not. The pendulum phenomena exposed when children mess about never fails to humble the teacher—or the highly trained scientist who has joined the fray.

Swing pendulums side by side. Make the bob go in a circle. Use a long stick for a pendulum or a baseball bat. Swing a pendulum underwater. Put a pendulum on a pulley and raise it while it swings. Imagine swinging a grandfather clock as a pendulum (thought experiment only)—it has a pendulum inside to keep time. Will a swinging grandfather clock run fast, slow, or not at all? Maybe swinging a pocket watch is a possibility.[8]

Before designing an investigation, children need to share descriptions and questions generated by simply messing about. What catches their attention and holds their fascination? Make these phenomena happen again for others to see and talk about in their own words as part of an "Exploration Symposium."

Messing about is prerequisite to being able to focus attention on what seems puzzling and in need of explanation. The place to share findings about pendulum motion is at a "Discovery Symposium." At this second symposium classmates can recreate features of their investigations, explain their designs, and then debate and question each other about measurements and variables.[9]

Theorizing—and debating—supersede arriving at conclusions congruent with conventional wisdom. What textbooks typically teach—changing length changes period—is not completely accurate, just a remarkably close approximation for small angles of bob displacement.

A sensitive pendulum with length and all other attributes kept constant will vary its period ever so slightly in response to variation in the earth's gravitational field. Properly calibrated, a pendulum can measure gravity—a laboratory exercise sometimes pursued in a college-level earth science course. Such an exercise suggests the value of continuity in teaching and revisiting pendulum motion as increasingly sophisticated understanding of physics develops.[10]

Farting Around with Computers

The continued sounds of snickering and giggling coming from the den concerned the two boys' parents. "They must be up to no good!" said Mrs. Green. "We'd better check on them."

They entered the den to find Felix, and his friend, Robbie, happily recording "bathroom sounds" with a microphone connected to the family computer. They had tired of playing Doom and felt mischievous. For these two middle schoolers, farting around with a computer was irresistible. Robbie's new oscilloscope app displayed the recordings on the screen as waves and made changing their frequency easy. They could cut and paste pieces of the pattern and repeat segments, too.

The two friends had made an interesting discovery about flatulence: compressing the waves created the high pitch calls of chirping birds. Stretching them turned their recordings into something that resembled the songs of humpback whales. "Well, maybe there's nothing wrong with what they're doing," said Mrs. Green to Mr. Green. "Just keep a window open," she advised the kids.

Apparently, the scope of messing about has expanded to include computer programs. Felix became a paleoclimatology scientist and climate modeler. As an adult, Felix analyzes global temperature data, obtained from isotope ratios from ocean floor sediments and ice core samples. He visualizes the data as oscillations from decadal to millennial timescales. Felix depends upon advanced statistical methods to tease apart trends, cycles, and random spikes.

Connect the dots of his interests and they lead back to adolescent computer play, at least in some small measure.

Shimmering Rainbows

Interest is curiosity about experience sustained over time. It begins with playfulness, works through puzzlement, and depends upon ample opportunities for exploration. Interest seeks to experience the unexpected, thrives on astonishment, and turns each new insight into several more questions. Interest is basic to learning science. Interest finds expression in questions and responds to an answer with another question.

One day Lorena, a first grader, came home from school excited about making rainbows with prisms. She had experienced prism phenomena in class as both an art lesson in colors and a science lesson in weather. This initial experience had piqued Lorena's curiosity and the time for serious messing about—for purposeful, playful exploration—had arrived. Lacking a prism but remembering how to use a mirror and dish of water as a substitute, her father sat down with her in the sunlight on the front porch. He held a sheet of white paper and told her how to tilt the mirror underwater while aiming it toward the sun. She then tried to aim the reflected beam at the paper.

Once Lorena understood the mechanics of the bowl, mirror, paper, sunlight system, her father left her alone, but continued to observe her from the nearby kitchen window. Quickly she became entranced with the shimmering spectrum of colors cast on the white paper. It was not quite like the one at school made with a rigid, plastic prism. She jiggled the bowl and noticed how the spectrum danced even more wildly. She tried to tilt the mirror and change the shape of the bands. She took the paper away and checked the colors against the blue wall of the house. The bands were straight, not curved like a rainbow, but this observation did not stir her interest. What did capture her attention was the water, first, in how it moved and second, in its clarity.

Lorena played purposefully with this system for about twenty minutes, then came inside with a request and a "telling question": "Can I have some food colorings? I can make a little rainbow with the mirror and bowl of water, but now I want to know, what will happen to those colors if I use colored water in the bowl?"

In this story the teacher designed an environment at school linking art and science in a sensory rich experience. The child came home to find support in her new interest and quickly put to use simple materials. An adult helped the child master the "mechanics" of the simple apparatus to ensure competency and avoid frustration. The child worked alone for a while, exploring several phenomena of interest, and gaining control over their occurrence. "Messing about" finally prompted a line of inquiry—changing the watercolor—captured in a "telling question." The story ended with the child sharing her findings with family and friends—including her teacher.

What is missing from this episode? Telling answers. There is no explanation of how spectra form from the dispersal of white light, why light refracts and reflects at the boundary of water and air, how a bowl of liquid can act like the triangular shaped prism, or what makes a rainbow curve. "Why" questions are not discouraged, just not tackled prematurely, as counseled in 1762 by Jean-Jacques Rousseau and David Hawkins in the 1965. Detailed answering can always wait if progress in asking good questions stays alive—if interest renders experience sufficiently intelligible, if control over inquiry helps focus attention on small dilemmas, potential anomalies, and describable incongruities.

The dancing motion of the spectra, the straightness of the rainbow bands, and the effect of colored water (or colored background) on the range of colors in the spectra charmed little Lorena. A scientifically satisfactory explanation of each would tax a college physics student. Providing the correct explanation to the child is not the function of the elementary teacher; it cannot be. Of much greater importance is helping the child acquire the habit and disposition of posing and pursuing enchanting questions. Creating interest matters most, and interest appears in the guise of questions. On interest, on questioning, all else depends.

For Lorena's class the study of rainbow color began as an art project, a reminder that children's science is "a-disciplinary" and need not represent knowledge in a particular subject. It's engagement that counts—having an interest, achieving success in its pursuit, and experiencing a sense of satisfaction.

ECHOES OF MESSING ABOUT

Loris Malaguzzi, architect of Reggio Emilia schools, judged the work of Francis and David Hawkins fundamental to teachers everywhere. For the past fifteen years the philosophy of David and Frances Hawkins has made its way into schools and museums in the form of an exhibit: *Cultivate the Scientist in Every Child*.[11]

The assembly of materials to follow up on messing about—and return to a messing about phase as needed—is crucial to the construction of a concept in a student's mind. As phrased in 1967 by Robert Karplus and Herbert D. Thier in the NSF's Science Curriculum Improvement Study (SCIS), a key component of the SCIS Learning Cycle is "conceptual invention." The other stages of the SCIS Learning Cycle echo Hawkins as well: exploration of a concept and concept application. In the 1980s, in the wake of the Learning Cycle's success, Rodger Bybee crafted a more elaborate cycle for the Biological Sciences Curriculum Study (BSCS), the 5E model: Engage, Explore, Explain, Elaborate, Evaluate.[12]

When strangeness dominates and piques curiosity, when prior knowledge feels inadequate for interpreting novel experience, when the previously familiar has turned unfamiliar because of a playful question, the time has come to talk, tinker, mess about, and talk, tinker, and mess about—and question—more. The engine of insight is creative imagination. Its gears are playful explorations with language and thought, things and actions.

Messing about is purposeful play with room for serendipity to strike—a truism at any age. Messing about, done well, sharpens perceptions, reveals anomalous events, discovers patterns, prompts wonder, and brings to awareness interesting phenomena to investigate further. Exploration grounds questions in need of answering through disciplined inquiry. An exploratory phase of learning readies even the minds of older students. Messing about may begin with whimsy, but it ends in serious work.

NOTES

1. Rudyard Kipling, "The Elephant's Child," in *Just So Stories* (Garden City, NY: Doubleday and Company, 1912), 63–64.

2. Jos Elstgeest, "Encounter, Interaction, Dialogue," in *Primary Science: Taking the Plunge*, ed. Wynne Harlen (London: Heinemann Educational Books, 1985).

3. David Hawkins, "Messing About in Science," *Science and Children* 2, no. 5 (February 1965): 5–9.

4. Karen Gallas, *Talking Their Way into Science* (New York: Teachers College Press, 1995).

5. Jody S. Hall, Carol Callahan, Helen Kitchel, Patricia Pierce, and Pedie O'Brien, *Organizing Wonder: Making Inquiry Science Work in the Elementary School* (Portsmouth, NH: Heinemann, 1998).

6. Elementary Science Study (ESS), *Teacher's Guide for Pendulums* (New York: Webster Division, McGraw-Hill, 1976). Originally published in 1966 by the Education Development Center.

7. Jearl Walker, "Strange Things Happen When Two Pendulums Interact through a Variety of Interconnections," *Scientific American* 253, no. 4, (October 1986): 176–180.

8. Jearl Walker, *The Flying Circus of Physics with Answers* (New York: John Wiley & Sons, 1977), 40, 237.

9. Jody S. Hall, *Organizing Wonder: Making Inquiry Science Work in the Elementary Classroom* (Portsmouth, NH: 1998).

10. Michael R. Matthews, *Time for Science Education: How Teaching the History and Philosophy of Pendulum Motion Can Contribute to Science Literacy* (New York: Kluwer Academic/Plenum, 2000). Matthews's treatment of the pendulum in the history of science offers opportunities for teaching much of physical science in the context of this story.

11. Alexander Cruikshank, "Messing About," exhibit component of *Cultivate the Scientist in Every Child: The Philosophy of David and Frances Hawkins*, accessed May 15, 2020, http://www.hawkinscenters.org/messing-about.html.

12. Rodger W. Bybee, Joseph A. Taylor, April Gardner, Pamela Van Scotter, Janet Carlson Powell, Anne Westbrook, and Nancy Landes, "The BSCS 5E Instructional Model: Origins and Effectiveness," *A Report Prepared for the Office of Science Education National Institutes of Health* (2006), accessed May 15, 2020, https://bscs.org/bscs-5e-instructional-model/.

Chapter 9

Salted Alcohol

The graduate students sat at tables with partners, sharing three transparent, numbered, plastic cups of clear liquids. Directions, presented in the spirit of playful exploration, were simple: "First, mark the level of the liquid in each cup. Next take an ice cube from the cooler and add it to the first cup. Share what happens." To no one's surprise, the ice floated and the liquid—presumably, water (tasting prohibited!)—rose.

"Now mark the new level. Next, guess what will happen to the level of the liquid when you place an ice cube in the second cup." Doing so only took moments and again the level of the liquid rose, but not as much as in cup number one. Hmmm . . .

"One more time, drop an ice cube into cup number three. Be careful not to splash or spill any of the liquid." A second or two later the room was buzzing. The ice had plunged to the bottom of the cup! "Look closely. What happened to the level of the liquid in the third cup? Did it rise more than, the same as, or less than in the first two cups?" The liquid in cup number three rose highest of all. Again, hmmm . . . Meanwhile, the ice in cup number two appeared to be melting more than the ice in the other two cups.

SINKING ICE LESSON

The "Sinking Ice Lesson" began with attention focused on a surprising event and then moved on to formalizing the meaning of "buoyancy." These were not elementary or middle school students. Instead, they were graduate students—mostly PhD candidates in linguistics (plus some master's candidates in geography, fine arts, and neuroscience)—at Beijing Normal University in the People's Republic of China. They had chosen to start their fall semester by enrolling in a workshop on "learning how to learn" organized by a quantum physicist, Wu Jinshan,[1] who was inspired by the work of Joseph Novak.[2]

The identities of the three liquids were easy to figure out: water, salty water, and isopropyl alcohol. The alcohol was detectable by odor. The three situations created two puzzles: (a) different levels of liquid in each cup after adding the ice and (b) the ice sinking in alcohol, but not in water. The density of the ice remained constant. The density of the salty water was highest and of the alcohol, lowest. Did the buoyancy of the ice—the force lifting it up against the downward force of its weight—vary, cup by cup, or had it remained the same?

Archimedes' Principle

Archimedes apparently was bathing when he noticed the telltale rise in the water level in his tub (maybe to where it left a noticeable ring). He linked this change to the buoyant force on his body, then shouted "Eureka!" The textbook, citing Archimedes' principle, explains, "A body wholly or partially submerged in a fluid is buoyed up by a force equal to the weight of the displaced fluid."[3]

Pressure increases as water deepens—ears can easily detect the increase in pressure from diving ten-feet deep into a pool. Pressure increases with depth at a rate that depends upon density. At equal depths, a column of salty water exerts more pressure than a column of freshwater. A column of alcohol at that same depth exerts even less. The pressure difference between the bottom and top is not enough to float ice.

Ice above the surface displaces no water; only the submerged ice does. A floating object displaces its own weight in water. A ship laden with steel pipes sits in a hole of water. The water to fill that hole weighs the same as the ship carrying steel pipe. The buoyant effect of a fluid works whether submergence is partial or full. The weight of the displaced volume equals the buoyant force.

The change in the level of the liquids in the cups recorded the volume of displaced liquid. Eyeballing the rise, in effect, roughly measured the displaced volume. The rise in the two cups of water measures the displaced volume needed to match the *weight* of the ice cube. In alcohol, the ice sinks, and the *volume* of the ice takes up space at the bottom of the cup. The ice cube's full size, not its weight (a force defined as mass × acceleration due to gravity), raises the level of the alcohol.

The class busied itself observing familiar and anomalous effects of density differences. Ordered from low to high, the densities in grams per cubic centimeter were .79 alcohol, .92 ice, 1.00 freshwater, 1.2 saturated saltwater. Precise measurement and calculation, however, were not the objectives of the activity; "messing about" held higher status. Careful observations of ice in the three cups of liquids led to the same ordering even without knowledge of precise values for density.

Playful Thought

Rather than starting with a definition of buoyancy or a statement of Archimedes' principle, the instruction crafted shared experiences and supported playful expression. Everyone understood that a brick underwater is easy to lift because it "weighs" less. A rock or brick is even easier to lift from the bottom of a saltwater pool. It feels lighter when submerged in alcohol, but the percent of its weight "lost" is less than percent lost in a denser fluid—where the buoyant force of lift is stronger.

Given this prior knowledge, a geologist once decided to carry a bucket of water to his banded iron formation field site. The geologist broke large fragments of dense rock to carry back to the laboratory and to share with a museum. Into the bucket of water went the samples. "Underwater the rocks weigh less and will therefore be easier to carry," reasoned the geologist.

The geologist, however, arrived back at his camp extremely tired, unable to understand why his load seemed so heavy. Beijing Normal University students had some excellent advice for the bucket-burdened geologist's next collecting foray into the field: "forget the bucket of water."[4] The rocks, after all, added their entire weight to the already hefty weight of the bucket of water. Humor, masked as a thought experiment, brought the first phase of the Sinking Ice Lesson to a close.

From Playful Exploration to Disciplined Thinking

The second phase of the buoyancy introduced several questions: "Why did the rise in the levels of the liquid in the three cups differ? Why did the ice sink in one? What happens to the level of the liquid in each cup as the ice melts?[5] What is the weight of an amount of water equal to the increase of the volume of liquid in the two cups of water? How might you figure out the strength of the buoyant force in each case? How much does the ice weigh when at the bottom of the cup of alcohol? What percent of its weight does an ice cube 'lose' when floating (100%)? What proportion of the ice cube remains above water?"

Nudged by the instructor, students considered the list of questions, picked one, and worked to agree on an answer argued from evidence. Everyone was welcome to search for helpful information. As the discussion played out, a few students looked up definitions using Baidu, the Chinese equivalent to Google. Some recalled the classical physics formula for calculating pressure: $P = \rho gh$.[6]

The students were encouraged to frame their thoughts in terms of gravity, density, pressure, buoyancy, and weight while challenging each other to share and clarify the meanings of these concepts. The exercise stressed the importance of expressing the effects of buoyancy in multiple ways:

- The liquid rose in each cup by a different amount: least in the salty water, most in the alcohol.
- Water in the first two cups rose as if small volumes of water had been added to the cups. That volume was—and had to be—the same as the portion of ice below the surface.
- The cube displaced a different volume of water in each cup.
- The proportion of the ice cube's volume floating above the surface was greater in freshwater than in saltwater. Zero percent of the ice protruded above the surface of the alcohol.

Changing the Density of a Liquid

Fascinated by the anomaly of sinking ice, two teams sought to find out the effect of adding salt to alcohol on the buoyancy of ice. Their creativity inspired an investigation of "salted alcohol" that proved interesting to everyone even though "salted alcohol" did not refer to an adult beverage.

Adding salt to alcohol first changed the "position of the ice" and then the "amount of ice above the water surface." After ten minutes the ice was in the middle of the cup. After twenty minutes the ice was at the top of the cup.

Lijuan and Yu Yan concluded, "Ice sink[s] in alcohol but float[s] in salted alcohol."

"As the salt melts in alcohol, the ice will float," explained Min, "the weight of displaced alcohol increases. Ice float[s] higher in salted alcohol because buoyant force equals to weight of displaced liquid compare[d] to weight of object."[7] The investigation successfully demonstrated what happens to buoyancy when salt is added to alcohol—"salted alcohol" in creative student jargon that maps experience into thinking.

FLOATING PLATE LESSONS

Learners never outgrow the need to mess about in science. The exploration of floating and sinking ice in water, saltwater, and alcohol gave tangible meaning to the concepts of density, buoyancy, and pressure. In subsequent classes, the Beijing students wondered about formations of ice, snow, and water as analogues to rocks. When the course turned to the theory of plate tectonics—and the question, "What moves the plates?"—density, buoyancy, and pressure proved useful.

Playful underpinnings give safe harbor to serious thoughts. Rocks, the same as ice in water and alcohol, float and sink according to density differences. The complexities of flow in the materials of the earth's interior build in large measure upon applying the concept of buoyancy. The astonishing effects include lifting the Tibetan Plateau and changing the world's climate as a result.

NOTES

1. Wu Jinshan, "Teach Less, Learn More at Being Normal University," 1st Asian Pacific Conference on Concept Mapping (CMC-AP 2017), Yingdong Conference Hall, Beijing Normal University (BNU), Beijing, People's Republic of China, September 20–22, 2017.

2. Joseph D. Novak, *Learning, Creating, and Using Knowledge: Concept Maps as Facilitative Tools in Schools and Corporations* (Mahwah, NJ: Lawrence Erlbaum Associates, 1998).

3. Paul A. Tipler, *Physics* (New York: Worth Publishers, Inc.), 125.

4. Rocks add their total weight to the bucket of water, so the geologist must lift both. Only when lifted in the water does the rock feel any lighter. Placed on a scale,

the bucket and the rock together weigh the same whether the rock is in the bucket or next to it. If the rock is placed on a submerged scale, it will register less weight. That's Archimedes' principle, and some common sense, in action. And not a half-bad "dad joke."

5. As floating ice melts, the water level does not change. In alcohol the cube occupies more space than its meltwater and the level of the liquid decreases by a small amount.

6. In this expression there is no value for area, volume, or weight, despite the definition of pressure as the ratio of force (weight force) to an area. *P* stands for pressure (measured in *pascals*), *h* for height (measured in *meters*), *g* for the acceleration due to gravity (measured in *meters per second, per second*), and the Greek symbol rho, ρ (pronounced "row"), stands for density (measured in *kilograms per cubic meter*). Pressure increases with depth. The rate depends upon the density of the fluid. More precisely, when a fluid is at rest—that is, there are no vertical accelerations within it—and held at a constant temperature, the pressure in that fluid increases as a function of the mass of the fluid above it. Pressure is the measure of weight over area, or mg/A (the definition of weight is mass times acceleration due to gravity, expressed as "mg"). Density is mass per volume, or m/V, an expression equivalent to m/A x h ("h" is the dimension of height in a volume). The expression density times volume, ρV, calculates total mass because mass adds up with an increase in volume at a rate dependent upon density. The expression for the total weight of a volume over an area becomes ρVg/A ("ρ," rho or density, is m/V—hence, "m" multiplies the "g," and thus weight appears in this expression). Now the trick is to simplify. Volume is height times width times length (h × w × l). The last two factors, w × l, equal area, A. So, V is h × A. The expression becomes: ρgAh/A. The A's cancel and ρgh stands alone, equal to P, the pressure below a column of fluid with height "h." That's the equation learned previously by the Beijing students.

7. The quotations from students were taken from written reports as written in English. Names have been changed.

Chapter 10

Squirrel Tales

Always seeking precision in his observations, Darwin made worm science the family business. He put his son Horace to work looking for worm castings in the cellar of a house. Francis set about collecting decomposing leaf fragments from worm burrows to test them for alkalinity.[1] William, a decade older than his male siblings, took on the task of recording whether worms dragged leaves into their burrows by the leaf stalk or leaf tip.

Darwin determined that his backyard earthworms paid no attention to the loud, low frequency notes played on a bassoon by Francis. "Shrill notes from a metal whistle" also did not interest Darwin's worms at all. Nor did wife Emma's piano music or shouts—that is, unless the piano was so close that the worm could feel the vibrations of the strings. Worms on a table ignored piano music from across the room.

Worms placed in a pot of soil *on* the piano, however, crawled at once into their burrows when Emma played C on the bass clef. When she struck G above the treble clef they similarly scurried into their burrows.[2] Darwin felt that this response made good sense from a worm's perspective. Worms sense vibrations transmitted through solid material. Certain frequencies may alert them to the digging of a predatory mole—perhaps the pitch range between bass clef C and treble clef G.[3,4]

Later in life he authored his final scholarly tome, *The Formation of Vegetable Mould through the Action of Worms, with Observations of Their Habits* (1881): a worm scientist at the ages of seven and seventy. His preschool collecting days disposed his mind to attend to nature's endless capacity to produce variation. His childhood encouraged him to experience delight and joy in the details of life. Disciplined learning transformed his playfulness into inquisitive habits of mind, even as it preserved his impertinence.[5]

NATURE STUDY

Anna Botsford Comstock received her diploma from Cornell University in 1885, just three years after Darwin's death. Her approach to education seemed to channel his spirit. Every backyard creature imaginable, from farm, field, and forest, seemed to find a page in her 1911 magnificent classic, *Handbook of Nature Study*, its lessons just as lively today as a century ago.

"In Nature-Study the work begins with any plant or creature which chances to interest the pupil."[6] Her *Handbook* includes a lengthy lesson on plant lice (aphids), an engaging one on pigeon wheat (an antiquated name for hair cap moss), and several on spiders and their types of webs. No one has surpassed her knack for framing questions readily answerable by direct observation—nor for experiencing intense delight in getting to know nature's common creatures, exactly the spirit that motivated the career of Charles Darwin and his magnum opus on garden worms.

Anna Botsford Comstock holds a prominent position in the history of women in science. Cornell University hired her first as a Nature Study assistant working with Liberty Hyde Bailey. In 1899, despite objections from some of the trustees, she became the first female professor at Cornell and achieved the rank of full professor in 1920.[7] Legions of graduate students in Science and Environmental Education at Cornell have sought opportunities to study at her rolltop desk, long kept in the graduate student office after her retirement. Many aspire to become experts at asking questions in the Nature Study style.

Bailey, a leading voice in agricultural research and world renown expert in palms, encouraged Comstock's work.[8] His *The Holy Earth* shared a theological perspective that made traditional religious leadership uncomfortable. In place of "salvation" as the cornerstone of Christian faith, he urged reverence for oneness in nature and the unity of living things, "the essential continuity of nature rising from within, rather than from without."[9] During the depression of 1891–1893 Bailey sent Comstock on a mission to restore the relationship between people and the land in rural New York—and thereby help to stem the tide of poor, rural families migrating to urban areas.[10]

Chickarees

When engaged in Nature Study, children watch and wonder about the doings of common creatures, often in their own backyards and gardens—for example, the behavior of a Chickaree (*Tamiasciurus douglasii*, also known as the Douglas squirrel or pine squirrel):

> Does the squirrel trot along or leap when running on the ground? Does it run straight ahead or stop at intervals for observation? When climbing a tree, does it go straight up, or move around the trunk? Does it go down the tree headfirst? How does it grip the limbs and tree trunk? In what direction do the toes point? Does the squirrel grip the bark with equal force with its hind and fore limbs when going up and down? How does it act when preparing to jump? How does it hold its legs and tail when in the air during a jump from branch to branch? Does the tail help it extend its jump or maneuver while in the air? How are the squirrel's eyes placed? Do you think it can see behind as well as in front all the time? Do its paws have claws? Is the tail as long as the body? Is it used to express emotion? Of what use is the tail when the squirrel is jumping? Of what use is it in the winter nest? Has it cheek-pouches like the chipmunk for carrying food? How does it express surprise, excitement, anger, or joy during the nut harvest? Note how many different sounds it makes and try to discover what they mean. How are its teeth arranged to gnaw holes in such hard substances as hickory nut shells?[11]

Nature study teachers learn to pose questions within the child's reach and to call upon the child's reasoning, in light of their observations, to reach answers. Comstock's questions, presented in 232 lessons, epitomize patience—the study of nature cannot be done in haste. There is no recipe for how to ask "Nature Study questions," but her *Handbook* has thousands of examples.

Through journaling and field study, Comstock's methods guide children's inquiry, connecting science to history, geography, gardening, language arts, and mathematics. In "Lesson 106: The Garden Snail," Comstock instructs teachers in how to build a "snailery." Prompted by rhyming couplets such as "With your eyes on stalks, and a foot that walks," children observe the snail climbing and clinging to the glass wall of the snailery and seek an answer to "Can a snail pull in one eye and leave the other one out?"

California Ground Squirrels

To pique interest, Comstock typically poses a "leading thought" early in the lesson. Often a dose of anomaly inoculates against disinterest. The information in a leading thought often works just as well, and sometimes better, as an end-of-lesson "Did you know . . ." question, for instance:

> Did you know that the California ground squirrel (*Otospermophilus beecheyi*) uses its bushy tail to deceive the Pacific rattlesnake? First the squirrel scatters loose dirt in the snake's direction. Disturbed, the snake shakes its rattle, the pitch indicating to the squirrel the snake's size and readiness to strike. With this information, the California ground squirrel begins to pump blood into its bushy tail and wave it back and forth. The rattlesnake sizes up its prey by sensing heat and takes the measure not of the squirrel's small body, but of the large fan of warm, wavy tail, and, unwilling to tangle with such a large creature, slithers away.[12]

Questions populate nature study lessons. Nature study questions are childlike, and nature closely observed holds their answers. The questions, scripted by the lesson or crafted by the teacher, suggest what to notice—and thus place success within reach for all children.

CHILD STUDY

Comstock's roots extend deeply into the history of educational reform. One of her mentors emerged directly from the Child Study Movement of the nineteenth century: Wilbur Jackman, dean of the School of Education at the University of Chicago, author of *Nature Study for the Common Schools* (1891). His work stressed the need for teachers to emphasize questions, not answers; no one exceeded Anna Botsford Comstock in this regard.[13]

Child Study applied the emergent field of educational psychology to the improvement of schooling—principally by acknowledging the centrality of interest and the changing nature of those interests as the child developed, stage by stage. Inheritors of this perspective included John Dewey and Jean Piaget.[14] Circa 1900, science teachers devoted themselves to laboratory methods with zeal even while recitation methods kept a hold on instruction in most contexts. Child Study, and its companion, Nature Study, sought to break this stranglehold.

Herbartian Philosophy

As a movement, Child Study proselytized the philosophy of early nineteenth-century Germany's Johann Friedrich Herbart. History credits Herbart with authorship of an enduring "theory of interest."[15] Interest, argued Herbart, developed from feeling pleasure from (a) "direct experience" with the natural world and (b) "social interactions" within a community. "Feeling is the bridge between cognition and volition," wrote Charles DeGarmo, author of *Herbart and the Herbartians* (1893). Understanding is a pleasurable, reinforcing emotion, and feeling the significance of an idea leads to acting accordingly.[16]

Verne Rockcastle, for decades steward of the Cornell Science Leaflets program, repeated Herbart's phrase "direct experience" countless times during his teaching career.[17] As recently as 2018, a Cornell cooperative extension pamphlet, "Trees: Dead or Alive," reiterated the twin Herbartian precepts of direct experience and social interaction:

> The world around us is animate, a source of deep wonder through observation and reflection, and through direct engagement with it, the basis of boundless insight. . . . This knowledge creates wonder and curiosity that act to widen a youth's perception and . . . create a wider sense of community. Book learning is not sufficient, as it must be accompanied by direct experience.[18]

Piagetian Merger

In the early 1960s, Rockcastle had the good fortune of spending time with Jean Piaget in France. He returned to Cornell and, with his colleague, Richard E. Ripple, organized a conference on cognitive studies and curriculum development. That event, in March 1964, introduced Piagetian psychology—and Jean Piaget in person—to American educators.

The conference report impacted the NSF's ESS and SCIS, ambitious new programs for teaching elementary science.[19] Note the significance of the word "study" in the titles of these projects. It signals their descent from the Child and Nature Study movements. Modern American science education merged Piaget and Herbart.

Herbart charged teachers with the responsibility of crafting experiences for children that challenged their curiosity. Anna Botsford Comstock brought his vision to fruition, as have decades of "study" movements that preceded standards-based instruction. Herbartians stressed the need for coherence among ideas—for each new idea to find "a place in relation to the already existing ideas,"[20] still a key principle of meaningful learning. Its realization rests upon masterful teachers' crafting the direct experience of simple things: squawking cups, funnel phones, twigs in glasses of water, saucers of salt evaporating on a windowsill, squirrels circling a tree, mirrors in a bowl reflecting sunlight, snails extending their eyes, and worms responding to the music of a bassoon.

NOTES

1. Charles R. Darwin, *The Formation of Vegetable Mould through the Action of Worms with Observations on Their Habits* (London: John Murray, 1881), 26–27, ed. John van Wyhe, *The Complete Work of Charles Darwin Online* (2002), http://darwin-online.org.uk/.
2. Ibid., 42.
3. Ibid., 27.
4. As reported in *The New York Times*, September 29, 2015 ("Turning the Worm Around," p. D4), neuroscientists at the Salk Institute successfully activated specific neurons in the nematode worm, *Caenorhabditis elegans*, using low pressure, ultrasound (low frequency) signals. The worm responded by making turns after being carefully sensitized (wild-type *C. elegans* made no such response). Stuart Ibsen, Ada Tong, Carolyn Schutt, Sadik Esener, and Sreekanth H. Chalasani, "Sonogenetics Is a Non-Invasive Approach to Activating Neurons in *Caenorhabditis Elegans*," *Nature Communications* 6, no. 8264 (15 September 2015), https://doi.org/10.1038/ncomms9264.
5. The section on Darwin and worms first appeared as "Irritating Worms," chap. 4 in *Do Elephants Have Knees? And Other Stories of Darwinian Origins*, ed. Charles R. Ault, Jr. (Comstock Associates/Cornell University Press, 2016), 56.
6. Anna Botsford Comstock, *Handbook of Nature Study* (Ithaca, NY: Comstock Associates/Cornell University Press, 1986), 5. Originally published in 1911.
7. Christy Peterson, "American Naturalists—Anna Botsford Comstock," *Kids Discover*, accessed May 15, 2020, https://www.kidsdiscover.com/teacherresources/american-naturalists-anna-botsford-comstock/.

8. Donna Eschenbrenner, "Then & Now: Philip Dorf's Biography of Liberty Hyde Bailey Gives Insight to Cornell Scientist," Special to *The Ithaca Journal* (October 18, 2019), accessed May 15, 2020, https://www.ithacajournal.com/story/news/local/2019/10/19/biography-profiles-pioneering-cornell-horticulture-professor-bailey/4006378002/.

9. Liberty Hyde Bailey, *The Holy Earth* (Ithaca, NY: New York State College of Agriculture and Life Sciences, 1980), 23. Originally published in 1915 by Scribner, New York.

10. Comstock, *Nature Study*, xi.

11. Ibid., 236–237.

12. Public Broadcasting System, "A Squirrel's Guide to Success," *Nature on PBS*, season 37, episode 4, accessed May 15, 2020, https://watch.opb.org/video/squirrels-guide-success-5nxkae/.

13. Audrey B. Champagne and Leopold E. Klopfer, "Pioneers of Elementary-School Science II: Anna Botsford Comstock," *Science Education* 63, no. 3 (1979): 299–322.

14. Herbert M. Kliebard, *The Struggle for the American Curriculum, 1893–1958* (Boston: Routledge & Kegan Paul, 1986).

15. George E. DeBoer, "Herbart and the Herbartians," in *A History of Ideas in Science Education: Implications for Practice* (New York: Teachers College Press, 1991), 24–30.

16. Charles DeGarmo, "Herbart, Johann Friedrich," in *The Encyclopedia Americana 14* (New York: Americana Corporation, 1953), 117. DeBoer's *A History of Ideas in Science Education*: *Implications for Practice*, ed. Lavinia P. Dudley (New York: Teachers College Press, 1991) draws upon Charles DeGarmo, *Herbart and the Herbartians* (New York: Charles Scribner's Sons, 1895).

17. In the interests of full disclosure, I must share that Verne Rockcastle chaired my doctoral committee at Cornell. My previous book, *Challenging Science Standards*, is dedicated to his memory. I first read an essay by Rockcastle while a student of Paul Kuerbis in Colorado prior to meeting Rockcastle. For both of their contributions to science education see Jon E. Pedersen, Kevin D. Finson, and Barbara S. Spector, *Going Back for Our Future II: Carrying Forward the Spirit of Pioneers of Science Education* (Charlotte, NC: Information Age Publishing, 2015).

18. Abigail Henderson and John Bowe Marianne E. Krasny, *New York's Natural Resources: Trees: Dead or Alive* (Ithaca, NY: Cornell University Cooperative Extension, October 15, 2018). Revision of Marianne E. Krasny, "Trees: Dead or Alive," 4-H Leader's Guide 147-L-22 (Ithaca, NY: Extension Publication of the Department of Natural Resources, New York State College of Agriculture and

Life Science, 1992). Krasny's work, in turn, was based upon Verne N. Rockcastle, *What's a Tree to Me?* (Washington, DC: National 4-H Council, 1981).

19. Edward L. Smith, Jane Heinze-Fry, Ellie Horwitz, and Don Duggan-Haas, "Verne Rockcastle: Life-Long Inquirer, Teacher, and Innovator," in Pedersen, Finson, and Spector, *Going Back for Our Future II*, 172–173. Smith et al., quote Ted Bredderman (a Rockcastle protégé) with regard to *Piaget Rediscovered* based on personal communication. Papers from the conference can be found in Richard E. Ripple and Verne N. Rockcastle, *Piaget Rediscovered: A Report of the Conference on Cognitive Studies and Curriculum Development* (Ithaca, NY: Cornell University, 1972).

20. DeGarmo, "Herbart, Johann Friedrich," 117.

Part III

AESTHETIC EXPRESSION

Chapter 11

Skull Sockets

Indira worked with enthusiasm as the team continued to look closely at the four plastic fossil replicas of teeth and skulls on the table, clues to the evolution of the horse. They paid close attention to crown height, ridge patterns, the position of the eye socket, and body length as a multiple of skull size.

"Ms. B., what's a hypocone?"

"I'm glad you asked, Indira. It's a cusp on a molar tooth, tongue-side, for grinding against the lower tooth."

"Thank you, Ms. B.," said Indira, using the class's affectionate name for their teacher. She rejoined her team and proceeded to answer the question in their lab exercise: "Do meat-eating mammals tend to have a hypocone on their teeth?" "No," they agreed, having found the answer in their notes: "Hypocones give teeth a square shape that increases surface area and helps plant-eaters grind their food."

Indira recalled fond memories of digging up broken bones on a New Mexico ranch and gluing them back together. At the age of eight, she imagined they came from an extinct beast, not a deceased steer. Handling the artifacts on the lab table reinforced her dream of one day becoming a paleontologist. True, the teeth and skulls were plastic replicas. Nevertheless, each one boasted an uncanny resemblance to the real thing, the exquisite details beckoning precise observations, a basis for inferring equid evolution.

The teams measured skull dimensions and observed tooth morphology for each specimen. Fossil hunters had exhumed the originals from sediments and volcanic ash spanning the past fifty-five million years of earth history (most of the Cenozoic Era). During that time, the Pacific and North American tectonic plates encroached upon each other. Their collision dramatically altered the landscapes and climates of the western United States.[1] (Movement continues in the present along the San Andreas fault and Cascadia subduction zone.) For Indira, the study of horse fossils was a field day.

STORIES FOSSILS TELL

Using dates as page numbers, the fossils told an evolutionary story set in North America. Populations changed as climate changed, and climate changed for two reasons: the reshaping of geography and the cycling of ice ages. Continental glaciations, both an effect and a cause of climate trends, opened and closed routes of migration. Folds and volcanoes, marking the movements of plates, aligned along a north to south axis, directly in the path of westerly winds from over the ocean. As a consequence, the aridity of the west increased, especially during the last two dozen or so millions of years.

Animal and plant adaptations tracked the shifting climate. Arid savannahs displaced lush forests; herbivore diets switched from twigs to grasses. Ancestors of the horse family scampered out of the woodlands, but the grasses of the expanding savannahs, laden with silica, abraded their teeth. Populations adapted or perished as natural selection rolled the dice. Some migrated. Fortuitously, larger variants thrived, trending to horses, asses, and zebras, the present-day survivors of an ancient lineage of odd-toed ungulates.

What traits shared by all of the specimens represented members of the horse family? What features of teeth and skulls varied from one fossil to another? How had shifting habitats refashioned the skulls and teeth, not to mention their limbs and toes, of the horse family members over tens of millions of years?

Some of the creatures probably had browsed; others had grazed. In the end, the size and placement of sockets in the skull, matched to body size, told a compelling story—the resizing, reshaping, and reforming of cusps and ridges of molar teeth accompanied by modifications in the shape of the head to accommodate them. Chewing silica rich plants wore down crowns; the evolution of really high-crowned teeth ("very hypsodont") met this challenge. Talking horses are rare, but their mouths have much to say from a Darwinian perspective. Tooth sockets deepened and eye sockets shifted to make room. Horse heads lengthened.

THE HORSE FOSSIL STUDY KIT

No one holds a plastic fossil on day one. Instead, Ms. B. distributes illustrations of the horse's foot bones and tree of evolution. The first is a detailed drawing in black and white of the front foot of a modern horse. Ms. B. directs everyone's attention to the upper bone (the metacarpal or "cannon" bone). Two "splint bones" grip its side, greatly reduced sibling metacarpals. Their image is distinct, but subtle. She reads, "These bones are sensitive and fragile, easily broken. They are of no use to a horse." The one to the outside has the same relative position as the hand bone boxers sometimes break.

The Mystery of the Splint Bones

The students find the bones in the drawing as Ms. B. asks, "Why do you think horses have these bones?" Pondering the anomaly of these supposedly useless bones introduces the story of equid evolution: how horses came to run on the nails at the tips of their middle fingers. Two other toes make a start, but horse genetics suppresses their growth. Why do horses have these splint bones? The teacher guide is mostly mistaken; the splint bones (fragile metacarpals) do bear weight. Added strength protects the cannon bones—equivalent (and, in the language of evolutionary biology, "homologous") to the middle bones of the hand in a human.[2] Clearly, after two decades of use, the Horse Fossil Study Kit would profit from some revision, yet the puzzle for starting the unit remains valid: What has happened to the horse's foot through tens of millions of years?

Bulbous Branches of Hippuses

The first illustration engaged the class in comparing bones of the human hand to the bones of the horse foot, one by one, from joint to joint, ending in the same tough stuff of fingernails and hooves, keratin (a protein). The second illustration (figure 11.1) introduced by Ms. B. was a chart of branching bulges littered with dozens of names ending in "hippus, hipparion, or therium"—but

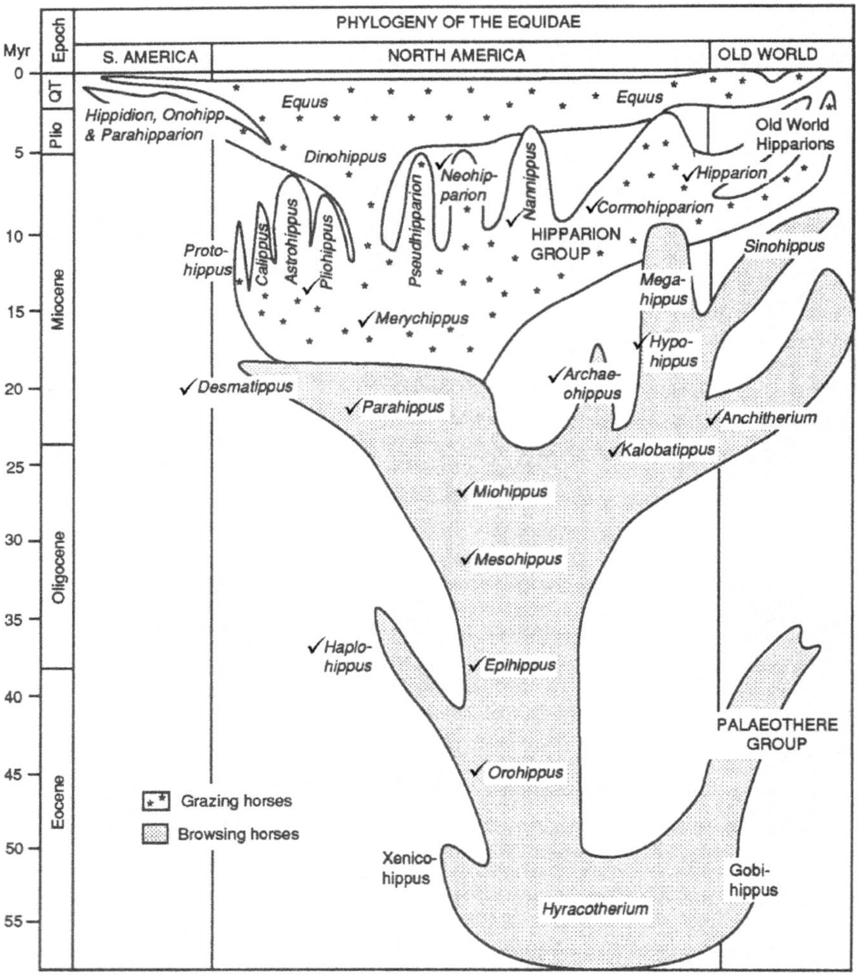

Figure 11.1 Chart of Horse Family Genera through Time by Geographic Location of Extinct and Living Genera. The chart illustrates migrations of ancient horses from North America. Check marks indicate genera found at the John Day Fossil Beds. *Source*: Illustration modified from *Fossil Horses* by Bruce J. MacFadden, ©1992. Reproduced with permission of the Press Syndicate of the University of Cambridge through PLSclear.

mostly "hippus" (the ancient Greek root is helpful here: "hippus" means "horse").

The names of continents appear as the top row. The vertical axis has a timescale in "millions of years ago" divided into epochs (Oligocene, Miocene, etc.). Shading separates the branched bulges into "grazing horses" and "browsing horses." The bulges seem to crawl up the page, through time, like the advancing pseudopods of an amoeba, spreading over the earth—ultimately, in search of grass.

The chart is a relationship tree, or phylogeny of the Equidae (horses and their close relatives). It charts their evolutionary descent as recorded in reliably dated fossils from around the globe.[3]

Given close scrutiny, the amoeboid blobs, set between axes of time and place, offer clues about the evolution of horses. "When, and for how long, did grazers and browsers co-exist?" asks Ms. B. (from about twenty to eight million years ago). "Where did grazers live?" (first in North America, then the Old World, and, eventually, South America). "Have there always been horses in the Americas since their origin?" asks Ms. B. The answer requires close inspection of the diagram.

Several branchings from the horse family placed evolving populations in Eurasia and Africa (the continents of the Old World) throughout the Cenozoic Era. They first crossed land bridges connecting North America to Eurasia via Greenland in Eocene time.

Most of the family's history unfolded in North America. After a robust and continuous presence for more than sixty million years, genera of modern horses became extinct in the land of their origin. Near the top of the diagram there is a subtle blank band indicating this disappearance. Ice Age climate and maybe predator pressure—including human hunters—decimated North American populations. Species of *Equus*, along with their compatriots *Megatherium*, *Mastodon* and much of the rest of the Pleistocene Megafauna, hung on until approximately 10,000 years ago—maybe even a bit longer—before succumbing to extinction on the New World stage. Members of the genus *Equus* that migrated across the Beringia Land Bridge from Alaska to Asia thrived in the Old World.

Darwin's Fossil Horse Tooth

A keen eye can discern from the graph that horses made two entries into South America. Primitive horses trotted south as the Isthmus of Panama formed nearly three million years ago. *Equus* (the modern genus) appeared in South America after a million-year gap in the fossil record via the same route. By eight thousand years ago, this genus was extinct in both of the Americas.

A curious, twenty-something Charles Darwin was the first European naturalist to find a horse fossil tooth in South America. In his journal he wrote, "In

North America bones of horses have been found in close proximity to those of the Mastodon; and I, at Santa Fe Bajada, found a horse's tooth in the same bank with parts of a Megatherium."[4]

The tooth crumbled out of the same bank in Punta Alta, Argentina, where he discovered the skeleton of a giant ground sloth, *Megatherium* (a life-size replica of this creature dominates the main exhibit hall at the Museo Carlos Darwin Museum in Punta Alta). The tooth came from an extinct South American species, *Equus curvidens*. Perhaps he shared his precious find with his Gaucho hosts while enjoying a supper of sautéed puma tongue and roast of armadillo. Horses, he realized, inhabited South America during Pleistocene time, then disappeared, only to return on board Spanish ships. In 1535 one carrying a cargo of seven stallions and five mares sailed into the harbor of Buenos Aires.[5]

Attention to the anomalous bone splints on the drawing of a horse's foot coupled to interpreting the diagram of equid phylogeny has primed the pump for horse fossil study. Before the replicas emerge from the traveling trunk, Ms. B. poses the essential question: "What does the geology and fossil record of the John Day Fossil Beds reveal about the climates and ecosystems of eastern Oregon's past?"

Films of Fossils in Clay

Ordered in time, the fossils of the John Day and Clarno Formation rocks of north central Oregon, especially the horse fossils, tell a compelling story of climate change. Film artists and field scientists have brought this story to life, and Ms. B.'s class appreciated learning science anchored to their own state.

The room darkens and the screen brightens as the film, *A Place of Discovery*, a virtual fieldtrip to John Day Country, begins.[6] Both animal and plant fossils abound. They tell of an epic journey through time in a land transformed from lush, tropical forest to arid grasslands, today's "high desert."[7]

The Cascade Mountains separate Oregon's rainy Willamette Valley from the rain-shadowed, eastern high desert. Almost eighty percent of Ms. B's biology class had never ventured out of the urban and suburban reaches of the valley. They were startled to think of Oregon as a land of juniper and prickly pear cactus, of badlands and rimrock—let alone a tropical, Panamanian jungle.

Long ago small "proto-horses" traipsed through swampy forests on splayed toes in this very same region. These little creatures were not really horses, but they receive credit for starting lineages that trended in that direction—and in another direction as well, leading to rhinoceroses.

Nothing astonished the class more than to think of horses and rhinos as relatives, except to learn that rhinos once roamed Oregon accompanied by other monstrous, yet quite unfamiliar beasts of the order Perissodactyla (Greek roots again: "odd" and "fingers"). Horse and rhino stock branched from the same limb, the limb being one that ended in an odd number of toes. They can count among their extinct cousins massive brontotheres, towering, clawed chalicotheres, and snouted tapirs. Rhinos and horses left the hemisphere; tapirs migrated south; the other odd-toed groups suffered extinction. (Horses from Spain, as noted previously, returned in the sixteenth century.)

A Place of Discovery scans the park's badlands to focus on paleontologist Ted Fremd as he prepares to chisel an oreodont skeleton out of its sepulcher. Soon Fremd is guiding the audience among the hills of the monument's Sheep Rock Unit (Oligocene Epoch), explaining the significance of these beds of ancient volcanic deposits to the task of puzzling out the Cenozoic history of western North America.

Even older rocks crop out in the Clarno Unit (Eocene Epoch). Modern streams have carved canyons through the slurries of mud, ash, and boulders that raced down the collapsing, water-soaked flanks of Eocene volcanoes. A trunk stands erect, the Hancock Tree, trapped in stone. Its tissues mark it as a relative of trees common from southeast India, across southern China, and down the Malay peninsula—lands wet and warm.[8]

Nearby the erosion of the modern landscape exposes the banks of a stream channel from a bygone epoch. Digging through channel sediments holds a surprise: exquisitely preserved nuts. Common are walnuts, moonseed, and palm. The "nut beds" reveal the secret history of plants with exuberance. Hundreds of species are represented, enough to reconstruct a detailed habitat.[9]

Mineralized stems of *Equisetum* grew along this stream, as do living *Equisetum* along nearby Bridge Creek. The name shares the genus name of horses: *Equus*. The common name of the plant is "horsetail." The segmented stems of horsetails branch bushy in the manner of a horse's tail. Ironically, horsetails are rich in silica—good for pot scrubbing—as are the grasses horses are happily able to graze, thanks to their hypsodont teeth's resistance to abrasion.

Even by Eocene time horsetails boasted an ancient lineage. They forested swamps in the Age of Coal, two hundred and seventy million years ago, prior to the existence of dinosaurs, let alone oreodonts and hippuses.

The film compresses fifty million years into twenty minutes. Formations of rock stack up as pages of a narrative; the last chapters, Mascall and Rattlesnake, end at the dawn of the Pliocene Epoch. By then grasslands dominate. Ms. B. summarizes the climatic shift starting with near-tropical, hardwood deciduous forests covering the Northwest, to the advent of grasses and fewer tropical trees, to tree and grass savannas, to grasslands, to the near-desert plants of the fossil bed's present landscape.

"What animals did you *not* expect to find living in the Northwest but were, in fact, preserved in the John Day Fossil Beds?" she asks.

"Camels." "Rhinos." "And saber-tooths!" Thoughts of truly giant beasts, bigger than today's biggest (whales excepted), enthrall the students. Bananas were not mentioned, but their fossils do turn up from time to time on John Day Country's slippery slopes.[10]

Ms. B. shares her story of visiting the monument one summer and hiking along the rim of Blue Basin where a friend spotted a flash of white that turned out to be a rhino fibula. "Did he keep it?" one student asked what everyone thought. "No," said Ms. B. "For a while it was displayed at the Cant Ranch museum, but a new exhibit replaced it with a more complete skeleton. The collection moved across the road into the Thomas Condon Paleontology Center near Picture Gorge. I think they keep the fibula in the research collection."

"The monument forbids collecting," she emphasizes. "You can't even pick up a piece of gravel from the parking lot."

In place of fossil *Equisetum* from the nut beds, Ms. B. next shares stems of modern horsetails. They circulate so that everyone can feel the coarse texture—the presence of silica figures later in the Horse Fossil Study Kit story line. "These plants still grow in wet soils across the state. They're easy to find."

The clock ticks and Ms. B. cues up *Impressions from the Past*, a film produced in 1987 in collaboration with professional artists and filmmakers and narrated by students of Dayville High School.[11] The combination of art and science provides a vicarious visit to the fossil beds. The film has withstood the test of time thanks to its segments of creative animation. Stop-motion, clay figures of oreodonts, nimravid saber-toothed cats, gracile rhinoceros, and three-toed horses star in the film. Lava oozes, trees topple, lahars tumble, land rifts, and climate changes as the peaceful valley of Dayville comes into existence.

A final, short film, *Equine Evolution*, zeroes in on the anatomy of members of the horse family through time. It begins with a drawing of the end points: little *Hyracotherium* from fifty-five million years standing under the belly of a modern horse. Drawings picture premolars and molars starkly contrasted with a black background. Premolars become larger and more complex through time. Bumps and ridges multiply. Animations of skeletons in motion focus attention on the elongation of the foot bones as horses increase in size and stride. Adaptations suitable for a diet of grasses appear at fifteen million years ago. From *Pliohipppus* to present, high-crowned teeth extend deep into their jawbone and skull sockets.[12]

Z Length

Ms. B. unveils the skeleton of a cat. "This skeleton is about the same size as the earliest known horse-like fossil, *Hyracotherium*." "No way!" chimes someone from the back of the room. Horse size increased through time and the head lengthened. The investigation starts with attention to head length.

"Each table has one or two fossil skulls," says Ms. B. "Choose one, then look at the diagram in your set of directions." The diagram depicts a horse skeleton with labeled sections of bones: neck length, head length, and body length. "Z" is the distance from the base of the neck at the top of the shoulders to the ground (see figure 11.2). "OK, everyone, go ahead and measure the length of the head."

Entire skeletons are fossil rarities compared to skull fragments and teeth. Knowing relationships that allow for scaling a full-size creature from a small piece has served paleontology very well. Students proceed to measure skull length as directed, then multiply this length by 2.4. The result is the approximate body length of the horse. "That's how a scientist estimates the full-size creature from the fossil," explains Ms. B.

Next, she presents a hypothesis: the sum of head length and neck length equals "Z" and never less than Z for *grazing* horses (Ms. B. adds the Z label to the diagram). "Why might this be so?" she asks. Ms. B. tells the class to picture lions hunting a giraffe and think about the Z hypothesis.

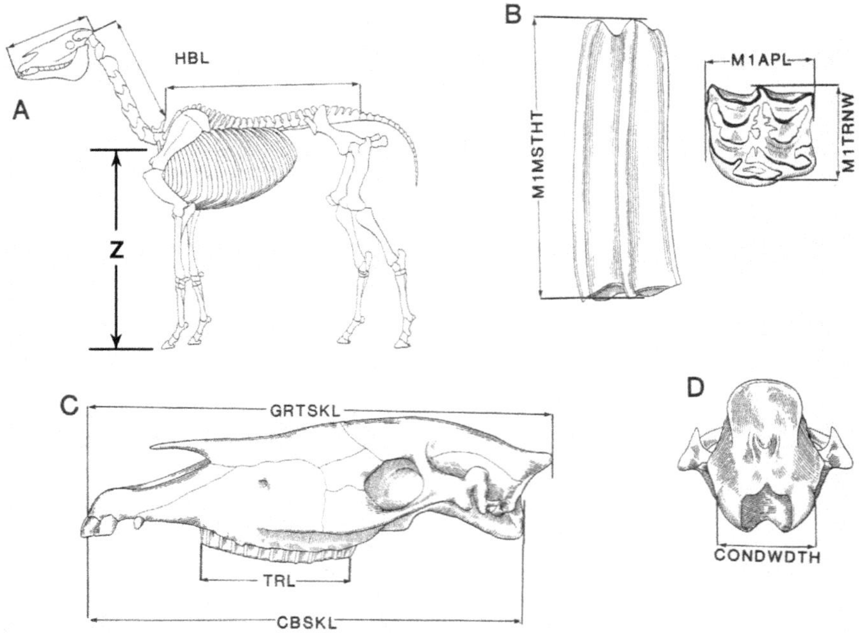

Figure 11.2 Skull and Teeth Measurements. Body Length (HBL) = 2.4 × Head Length (GRTSKL) for grazing horses. The molar tooth, drawn from both side and top down views has a high crown (very hypsodont). "Z" is the height above the ground of the joint at the base of the neck. *Source*: Illustration modified from *Fossil Horses* by Bruce J. MacFadden, ©1992. Reproduced with permission of the Press Syndicate of the University of Cambridge through PLSclear.

In short order a student volunteers an answer, "Maybe so they could stay standing with their mouths able to reach the ground."

"Or stand while taking a drink."

"Why would standing be a good way to eat or drink?"

"I know! To run away faster from a saber-tooth."

"Yes," Ms. B. adds. "Kneeling puts a horse in danger. Being ready to run can save its life."

"Browsers often lift their heads to eat. For browsers, the Z length can be less than the sum of head and neck length without causing a problem," Ms. B. explains.

The Anterior Orbit Point (AOP)

Once again students are guided in collecting visual observations. This time the data are the relative positions of eye sockets ("orbits") with respect to dentition. A team leader reads the directions: "Imagine a line straight down from the AOP (the front of the eye socket) that intersects the line of teeth at ninety degrees. In terms of the teeth, where is that intersection point?" Students refer to a drawing that labels the teeth on the upper palate: P1, P2, P3, P4, M1, M2, and M3 for premolars and molars, respectively. Specimen by specimen the students share answers as Ms. B. writes them down and projects them as a table for the class to copy.

The class looks at the table as Ms. B. asks, "In which skull is the AOP the most forward?" Numerals preceded by capital letters label each specimen. There are no names and no dates. Masking identity short circuits quick Googling for answers. Students must depend upon their data.

"In AMNH 4832. It's above the back half of the first molar (M1)," Indira answers with enthusiasm. AMNH 4832 is the replica of a *Hyracotherium* skull from fifty-five million years ago, a presumed browser.

Again, in reference to the AOP/dentition table, Ms. B. asks: "On which skull is the AOP farthest back?"

"LACM 1863. It's even behind the last molar (M3)," adds another student.

"That's right," Ms. B. confirms. "There are no teeth at all directly below the AOP on LACM 1863." The specimen, from the Los Angeles County Museum (LACM), is from a modern horse, *Equus*, a grazer. The AOP position with respect to teeth for all of the other specimens falls between the positions observed on *Hyracotherium* and *Equus*.

Of Time and Teeth

The class has learned from the films that "tooth patterns tell time." To review this information, Ms. B. shares two enlarged drawings of teeth: an ancient tooth with a crown of cones and a modern horse tooth with a convoluted, more complex pattern of ridges. For horses, the trend is an increase in the complexity of ridgelines joined to lengthening in the height of the crown, plus the addition between the enamel ridgeline of a hardening agent, "cementum" (it's also the substance that anchors teeth to sockets).

The class is well prepared to answer the focus question for the lab activity: "Given what you know about teeth patterns through time, and based upon your observations of the four specimens at your table, which one is likely the most ancient?" The teams reach consensus rather quickly: the lowest crowned, least ridged tooth (cementum cannot be distinguished from enamel in the plastic replicas).

The challenge gets harder. Some of the specimens are teeth embedded in the palate of the skull (maxillary bone); others are fragments of mandible (jawbone). The challenge continues until all four fossils in the table set are placed in a line from most ancient to most recent. Again, achieving consensus about "most recent" comes quickly. The middle specimens prove to be the most vexing, as they should. Intermediates happen.

Everyone rotates to a new table to repeat the task—four times in all. Sometimes the sequence is too close to call. But the end points are always obvious—to those with a trained eye.

After the last rotation, Ms. B. projects the image of a tooth viewed from the top of the crown for everyone to see. "Which specimen has a tooth that matches this one?" It's the oldest tooth in the collection—M1, right side, maxillary found at table three on specimen AMNH 4832—*Hyracotherium*, the proto-horse critter from fifty-five million years ago.

The Horse Family Tree

The class anticipates the grand finale: putting all sixteen fossils in order from most ancient to most recent. They have done so, in relative terms, based on tooth patterns from brachydont to hypsodont and skull morphology (skull length and position of the AOP) for each of the four sets.

Starting with one team's set of four fossils, Ms. B. has each team in turn carefully place their specimens on the table in the front center of the room. They may position each specimen as older, younger, or about the same age as the ones already present. They may reposition fossils already placed as they add ones for their table set, but they must offer evidence in support of the change (multiple interpretations are possible) The task gets harder and harder. The time line spreads wider to accommodate specimens from "about the same time." They complete a tree of descent and relationship extending from the past to the present, an evolutionary solution considered reasonable in light of their observations and the limited scope of their collections.

Unless dating is independent of the ordering pattern, temporal reasoning becomes circular. Volcanic sediments buried John Day Country fossils and volcanoes erupted independent of how creatures behave. Sediments bracketed with volcanic tuffs (ash and glass, settled from hot clouds) can be dated, explained by Ted Fremd in the first lesson's film, by "single crystal, argon-argon" methods to yield dates accurate to within one hundred thousand years—just one percent of ten million years. "Within one percent" is surprisingly good in paleontological work! Radiometric clocks tell John Day time rather reliably.

How well does the class's solution hold up when independent dates are introduced for the age of each fossil? To find out, the class returns to the diagram of the Phylogeny of the Equidae (figure 11.1), the bulbous, branching horse family tree set against a time line, shaded to indicate browser or grazer, and mapped onto continents. Up to this point, students have depended upon museum collection codes to identify and sort the specimens. None has a time stamp or genus label. The diagram features scientific names for different genera of horses, but not the collection codes stenciled on the specimens. Ms. B. has the matching key.

She begins to call out the genus name for each catalogued specimen. "JODA 3363: *Kalobatippus*" (JODA stands for John Day Fossil Beds National Monument).The names of thirty-one genera appear on the diagram of the Phylogeny of the Equidae; the Horse Fossil Study Kit contains sixteen specimens and includes duplicates, thus just eleven separate genera. It takes a few moments for the class to find the genus *Kalobatippus* on the diagram, positioned at twenty-four million years ago on the border between North America and the Old World.

"Write down JODA 3363 in small letters right beneath *Kalobatippus* on the diagram of equid phylogeny," Ms. B. instructs the class. In the spirit of a game of bingo, she encourages students to "call out 'hippus' when you have entered the names of all four of your specimens on the horse family tree!"

The investigation comes to an end with a return to the question that focused attention on the horse family tree before anyone had inspected the fossils. "Look at your horse family tree diagram. When did browsers and grazers live side by side on North America?"

Ms. B. continues to pose questions that call for interpretation of the data. "Which specimens have JODA labels? What is the range of ages for the JODA collections specimens in the Horse Fossil Study Kit?" The range spans about ten or twelve million years and crosses the boundary between Oligocene and Miocene Epochs, an especially interesting period of geologic and evolutionary history. The specimens, from oldest to youngest, are *Miohipppus*, *Kalobatipppus*, *Parahippus*, and *Merychippus*. More than a dozen other, older and younger specimens of Equidae have been unearthed at the John Day Fossil Beds, from the Eocene Clarno Formation, through the mostly Oligocene John Day Formation, and into the Miocene Mascall and Rattlesnake Formations.[13]

A plethora of mammals trotted through the Pleistocene of North America in the company of horses, each group with its own distinctive dentition, reflective of diet. To test their perceptual acuity and horse dentition savvy, Ms. B. projects a figure with black and white drawings of eight teeth from living mammals: peccary, kangaroo, warthog, deer, rhinoceros, horse, woodchuck, and the capybara (large rodent from South America). Five of the teeth are square-shaped (see figure 11.3). "Can you find the horse tooth? It's not the only one with a hypocone!"

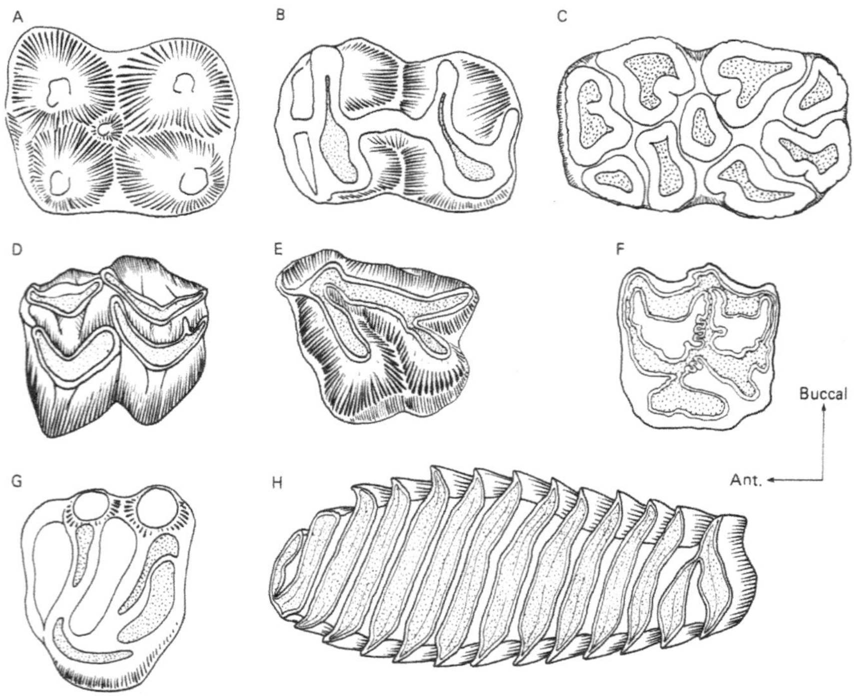

Figure 11.3 Drawings of Modern Mammal Molar Teeth Crowns. The horse tooth is on the right end of the second row. *Source*: Illustration from *Fossil Horses* by Bruce J. MacFadden, ©1992. Reproduced with permission of the Press Syndicate of the University of Cambridge through PLSclear.

Having spent nearly ninety-minutes sleuthing tooth fossils, everyone had no difficulty spotting the horse tooth (right end of the second row). And thus the first part of the class's vicarious journey to the John Day Fossil beds, courtesy of plastic replicas of skulls and teeth, came to an end.

NOTES

1. C. Page Chamberlain, Hari T. Mix, Andreas Mulch, Michael T. Hren, Malinda L. Kent-Corson, Steven J. Davis, Travis W. Horton, and Stephan A. Graham, "The Cenozoic Climatic and Topographic Evolution of the Western North American Cordillera," *American Journal of Science* 312 (February 2012): 213–262, https://doi.org/10.2475/02.2012.05.

2. Les Sellnow, "What's a Splint?" *The Horse: Your Guide to Equine Care* (October 24, 2013), https://thehorse.com/16474/whats-a-splint/.

3. Bruce J. MacFadden, *Fossil Horses: Systematics, Paleobiology, and Evolution of the Family Equidae* (Cambridge: Cambridge University Press, 1992).

4. Charles R. Darwin, *Charles Darwin's Beagle Diary*, ed. Richard Keynes (Cambridge: Cambridge University Press, 2001), 182.

5. Richard Darwin Keynes, *Fossils, Finches, and Fuegians: Darwin's Adventures and Discoveries on the Beagle* (New York: Oxford University Press, 2003), 199.

6. Oregonians refer to the north central region of their state as John Day Country in memory of a fur trader legend. In 1811 John Day and Ramsey Crooks, of John Jacob Astor's Pacific Fur Company, huddled on the shore of the Columbia River at the mouth of one of its tributaries. Rescued, they returned to Astoria—John Day, as the legend goes, in a state of madness, his name lingering on as an eponymous town, the nearby Dayville, the John Day Dam, the John Day River, and, of course, the John Day Fossil Beds National Monument. The original source for the story of John Day is Washington Irving's *Astoria, Or, Anecdotes of an Enterprise Beyond the Rocky Mountains* (Chicago: Belford Clarke & Co., 1884), 268, accessed June 10, 2020, https://www.questia.com/read/13826026/astoria-or-anecdotes-of-an-enterprise-beyond-the-rocky-mountains.

7. *A Place of Discovery*, directed by Anne Tubiolo, written by Anne Tubiolo and the John Day Fossil Beds National Monument staff, narrated by Mary Whittington (Washington, DC: U.S. Department of the Interior National Park Service, 2001).

8. Elisabeth A. Wheeler and Steven R. Manchester, "Middle Eocene Trees of the Clarno Petrified Forest, John Day Fossil Beds National Monument, Oregon," *PaleoBios* 30, no. 3 (April 18, 2024): 105–114.

9. Gregory J. Retallack, "A Field Guide to Mid-Tertiary Paleosols and Paleoclimatic Changes in the High Desert of Central Oregon—Part 1," *Oregon Geology* 53, no., 3 (May 1991): 51–59.

10. Mary Caperton Morton, "Finding Fossils in John Day," *Geotimes: Earth, Energy, and Environment News* (March, 2008), accessed June 10, 2020, http://www.geotimes.org/mar08/article.html?id=Travels0308.html.

11. *Impressions from the Past: The John Day Fossil Beds*, directed by Cynthia Tait and Jim Likowski, narrated by Kelly Christman, animated by Lonnie Smith and the students of Dayville High School, Cynthia Tait, Instructor (Portland, OR: Northwest Film & Video Center, 1987).

12. *Equine Evolution*, directed and animated by Peter Crosman (Washington, DC: National Museum of Natural History, 1985).

13. John P. Graham, *John Day Fossil Beds National Monument Geologic Resources Inventory Report Natural Resource Report* (Fort Collins, CO: U. S. Department of the Interior National Park Service, Natural Resource Stewardship and Science, 2014), accessed June 10, 2020, http://npshistory.com/publications/joda/nrr-2014-846.pdf.

Chapter 12

Crowned Molars

Packed away in their traveling trunk, the equid skulls rest peacefully. In the new week, a new challenge confronts the class. By now everyone realizes that the members of the horse family increased in body length as Cenozoic time unfolded. Teeth crowned higher and higher, too. Do the fossils record probable pathways connecting possible ancestors to presumed progeny, or are the samplings of extinct creatures too sparse to tell? Does the paleontological evidence justify the claim of evolution from browsers to grazers?

The secret to equid survival is to eat and not be eaten. To eat well, horse teeth must harvest plants in abundance. Plants, in turn, struggle to defend themselves. To resist predation, horse populations adapt their own defenses (size and lethal kicks) and escape strategies (a fast gallop). Predators, in size, strength, speed, fang, and claw, evolve accordingly or vanish. The evolutionary and extinction records of producers (plants), consumers (horses), and predators (big cats) are an ecological drama played out on a stage set by geologic processes.

Anatomical patterns are a starting point in the study of fossils. Megabeast reconstruction has its intrinsic rewards. There are, in addition, significant branches to follow in the browser-grazer story line. Fossils embed information in their microscale composition. Their distribution geographically and temporally tracks climate change. Lessons in the Horse Fossil Study Kit reach beyond the record of equid descent to delve into the active interplay among geology, evolution, ecology, and climate.

PENNIES FOR THOUGHTS

Drawing evolutionary inferences from paleontological data depends upon sampling, and sampling encompasses fossil ranges in both time and place. If finds are few, uncertainties are large. The paleontologist's gold standard is to build collections representative of geographic and temporal variation.

How many fossil discoveries are needed to warrant a claim of descent within the horse family? Lots. The teachers' mission, in the post-plastic replica lessons of the Horse Fossil Study Kit, is to deepen student engagement in thinking paleontological thoughts—to reason probabilistically and evaluate the robustness of sets of fossils from different locations and horizons in time.

Activity begins with the distribution of pennies, each symbolic of a fossil find, and a set of three charts (see figure 12.1, the Lab 3 Master, Paleo Team #1, and Paleo Teams #2 and 3 charts with dots for specimens). The charts depict body length on the horizontal axis and age of rock strata on the vertical

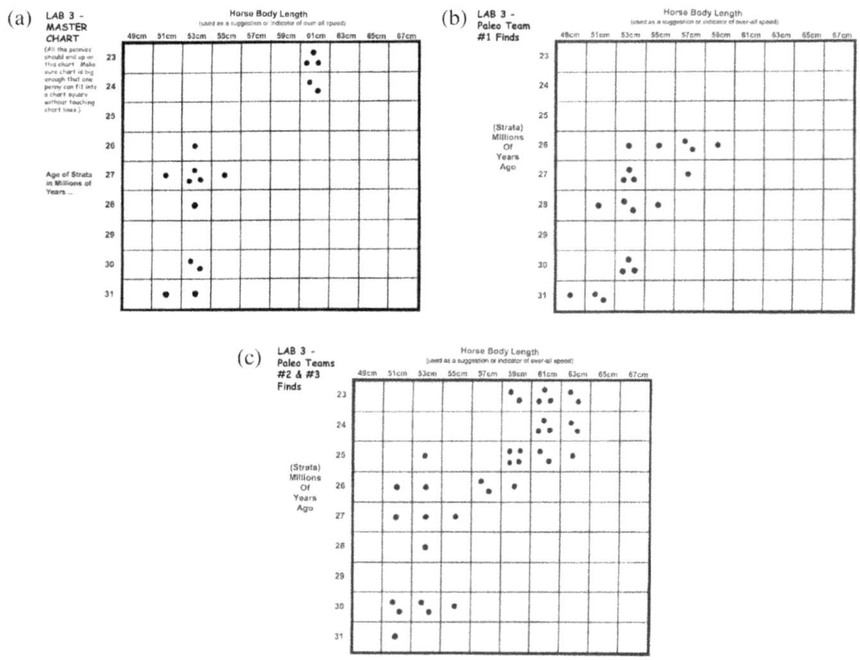

Figure 12.1 Three Charts for Plotting Equid Fossil Body Length Twenty-Three to Thirty-One million years ago from Lab 3 in the Horse Fossil Study Kit. By stacking pennies, students add to the samples of body sizes on the Master Chart with the findings of field research teams 1, 2, and 3. *Source:* Graphic by John Fiedor, Chief of Interpretation at John Day Fossil Beds. Reproduced by permission of John Day Fossil Beds National Monument.

one. On each chart Horse Fossil Body Length increases from left to right. The Age of Strata (the layers of rock where the fossils were discovered), decreases from bottom to top. Body length spans forty-seven to sixty-seven centimeters and age from thirty-one to twenty-three million years ago.

Starting with the Master Chart, students build a graph in bingo fashion by placing a penny in the proper box: "Row 28, column 53!" Ms. B. exclaims and students place their first penny.

The initial chart (the "master"), after the first round of data entry, has pennies stacked three high in just two boxes with a few scattered pennies around them. Most pennies cluster around two boxes. Rows and columns of blank boxes separate these clusters. A third cluster of four specimens appears at the bottom—the oldest fossils and shortest body lengths.

Does the distribution justify linking the three clusters as a pathway of evolution? It's tempting, but the point of the lesson is that the data are too sparse to draw this inference with confidence. Three-dimensional graphing lays a foundation for probabilistic reasoning without the introduction of statistical procedures. Even the visual pattern at this stage, however, does not justify inferring a pattern of relationship.

The activity, in simulation mode, proceeds. The teams imagine they have received support from the fictitious Paleolands Research Institute to organize a field expedition to the John Day Fossil Beds. In strata from thirty-one to twenty-six million years ago, the expedition finds twenty new specimens suitable for determining body length (figure 12.1, Paleo Team #1 chart). Each finding becomes a penny on the chart. Stacks of pennies continue to rise at the intersections of age and body length.

The stacks in the left-of-center, middle of the chart, cluster are now taller. Empty boxes, however, still separate the mid-chart horses from the cluster of small horses at the bottom (the oldest fossils). With the addition of the discoveries of four fossils indicating body lengths of fifty-seven to fifty-nine centimeters, the mid-chart cluster spreads to the right. The cluster seems to be encroaching on the territory held by the pennies representing larger horses.

Transition Fossils

Ms. B. shares with the class additional findings by researchers who have studied the John Day Fossil Beds collections. The class learns that horses with body lengths up to fifty-three centimeters nearly uniformly displayed the teeth patterns of browsers. Moreover, horses sixty-one centimeters in length, plus and minus two centimeters, all had teeth suitable for shearing grass. A trend in tooth pattern through time may be emerging from the fossil data.

Paleontological researchers report interesting changes in John Day Country climate twenty-seven to twenty-three million years ago: vast forests have receded, yielding to open savannahs and grassy plains. Does the distribution now justify linking the three clusters as a pathway of evolution?

A blank row at twenty-nine million years ago (figure 12.1 with entries from Paleo Teams #2 and #3 added) still isolates the cluster of oldest fossils of small, browsing horses. The trend to the large horses remains sparsely populated. The answer to the question remains "no."

How might additional discoveries contribute to solving the problem? The row for twenty-five million years ago continues to be blank. There are only three horse fossils on the chart in the cluster representing long body lengths.

The class has no data from a critical time period! What sizes of horses existed right in the middle of the time when forests declined and savannahs advanced? Discoveries from this horizon in time might connect the pennies and confirm the hypothesis of the descent of grazers from browsing ancestors.

The Paleolands Research Institute announces a new round of funding and, impressed by the Ms. B.'s class's savvy questioning, grants them additional funding for expeditions to the John Day Fossil Beds. Two teams embark to prospect for fossils in beds straddling twenty-five million years ago. They return toting crates of horse fossils wrapped in burlap and plaster *from a time of ecological transition.*

Adding specimens means stacking more pennies on the master chart. How convincing has the pattern become? In three dimensions the pennies form two hills. A ridge in the row for twenty-six million years ago connects their peaks and suggests a time of transition (figure 12.2 displays dots for pennies and does not capture the three-dimensional nature of the charted data). The samples of fossils making the ridge increase in body length. The ridge separates two peaks—easily observed as stacks of pennies. There is a peak for fifty-three-centimeter horses at twenty-seven million years ago and another rising for sixty-one-centimeter horses from twenty-four to twenty-three million years ago.

The smaller browsers, closing in size on the grazers, certainly look like their forbearers. No evidence has turned up for another population, at least in the John Day beds, for a different set of ancestral candidates. Migration

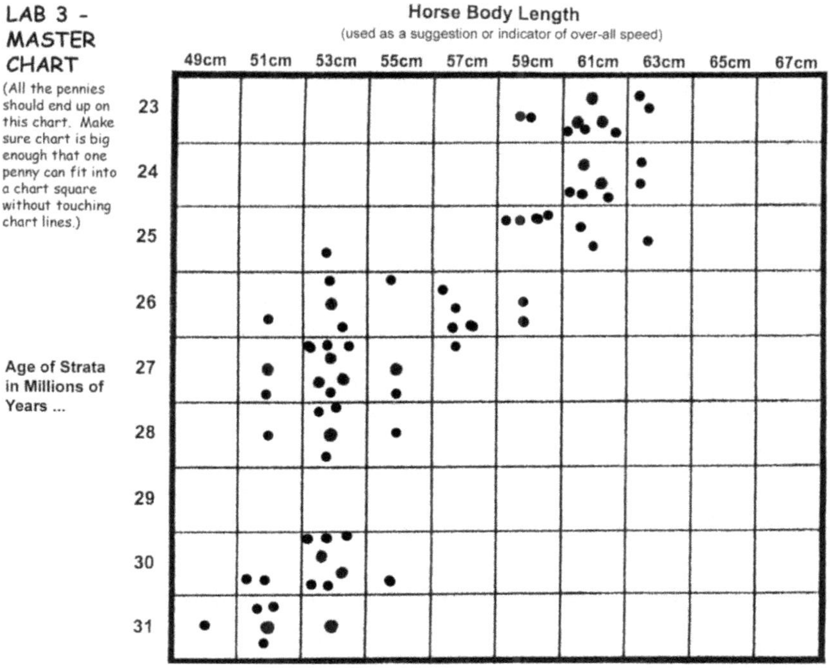

Figure 12.2 The Completed Scatter Plot on the Master Chart Correlating Horse Body Length with Age of Strata in Millions of Years Ago from Lab 3 in the Horse Fossil Study Kit. Stacking pennies, representing the discoveries of fossil specimens by all three field teams, forms two peaks connected by a ridge across the row for twenty-six million years ago. *Source*: Graphic by John Fiedor, Chief of Interpretation at John Day Fossil Beds. Reproduced by permission of the John Day Fossil Beds National Monument.

cannot be ruled out, but the research data from Wyoming's Wind River Basin and the Wasatch Mountains of Utah reinforces the pattern discovered in Oregon.[1] The evidence for a connection is compelling.

The transition at about twenty-five million years ago crosses the boundary in the John Day Fossil Beds between the geologic records of the Oligocene and Miocene Epochs. Paleontologists have quarried seven species of horses from strata near that boundary. The Horse Fossil Study Kit includes replicas of four of them: *Miohipppus*, *Kalobatipppus*, *Parahippus*, and *Merychippus*.

Ms. B. reopens the kit to take these replicas and gives a simple direction, "Look at your stack of pennies and the Phylogeny of the Equidae diagram (look again at figure 11.1, previous chapter). Which pennies might represent your specimen?" *Kalobatippus* seem to fall squarely on the twenty-five-million-years-ago row. Oligocene *Miohippus* lived a couple of millions of years earlier; *Parahippus*, two or three million years later in the Miocene. *Merychippus*, the youngest, trotted about near the fifteen million year ago mark of the Miocene, much too recent to claim a penny position on the graph.

How are the fossils similar? How are they different?" *Merychippus*, the youngest, falls into the category of "grazer." The other three had browser features. Revisiting the replicas in the context of interpreting the charts deepens their meaning along with insights into how this particular science gets done.

Predator Pressure

Natural selection is a blind opportunist. It converts unplanned variation into useful adaptations. In the struggle to survive, predator and prey must counter each other's shifting advantage. They never reach a truce, though they sometimes come close to achieving an equilibrium. Often the contest spirals into an "arms race," as shared in the final lessons of the Horse Fossil Study Kit.

When viewed from eye level, the three-dimensional stacks of pennies (the rudiments of statistical insights) reflect an unobvious signal in the fossil record—cat predation! Ms. B. introduces this scenario with the question, "Who do you suppose feasted on the small horses that browsed forested habitats?"

She asks them to picture the answer: nimravids, a stocky, saber-toothed, ambush predator that hunted in North America and Eurasia from thirty-one to twenty-five million years ago. Ms. B. adds that larger species of cats, saber-toothed felids, likely preyed on the larger, faster horses later in time. Felids made an entry in the fossil record at about twenty-seven million years ago. Both nimravids and felids mysteriously disappeared from North America at about twenty-four million years ago—and mysteriously returned four million years later.[2] In the meantime, saber-teeth pierced the necks of many horses.

What have the pennies to say about populations of ancient, saber-toothed cats? Little, if anything, about how they evolved, but something interesting about their extinctions. Small browsing horses, according to the stacks of pennies, disappeared in the John Day fossil record twenty-five million years ago. Nimravids disappeared from the fossil record in this region at about the same time. Coincidence? Probably not. Their simultaneous demise suggests something happened that upset the food chain. Climate shifted and the plant community followed. Food for browsers became scare and their population dwindled. As their prey declined, nimravids likely did, too.

A second story plays out in the graph between the rows for larger horses at twenty-four and twenty-three million years ago. The number of fossil horses before and after twenty-three million years ago gently rises and falls. At twenty-four million years ago, the curve shoots up sharply, then tails off slowly. A teacher cannot expect students to discern this difference without direct instruction, but once they have compared the steeply and gently sloping sides of the sampling distribution, they can speculate in answer to "Why does the curve—the climb up and down the hill—change shape?" Ms. B. directs them to a clue: the range in time of felids in the fossil record. Have horse fossil hunters detected the effects of predation by big cats, as today practiced by lions on zebras?

Perhaps the larger sabertooths (felids), by preying successfully on smaller variants of horses, pressured the overall horse population to trend larger. The range of horse sizes seems to have expanded after saber-toothed cats departed. The extinction of felids in North America may have relaxed predation pressure and allowed the range of body lengths among horses to return to its more normal distribution.

Being big helped to make horses safer. Being faster rode in company with being bigger. Eyes to the side of the skull, a feature of prey species from squirrels to zebras, caught glimpses of predators better than eyes focused straight ahead. Drinking and grazing on straight legs, rather than bent knees, aided quick getaways. Horses' wariness, speed, and size evolved in an arms race with agile, quick, saber-toothed, big cats. At the same time, another, more subtle, arms race was unfolding.

TEETH MADE OF GRASS

Changing climates from the Eocene through Pliocene Epochs altered seasonal patterns of rainfall and temperature, transforming floras. Climate changed, plants changed, plant-eaters changed: an evolutionary trifecta. Plants adapted to herbivore feeding habits; herbivores adapted to plant defenses and nutrient quality.[3]

Mouths unable to cope with emerging grasslands doomed some creatures, while variations trending toward abrasion-resistant teeth and elongated, flexible upper lips created opportunities for others. Horses became long in tooth as natural selection played out. In Darwin's words:

> We shall best understand the probable course of natural selection by taking the case of a country undergoing some physical change, for instance, of climate . . . and some species might become extinct. . . . In such case, every slight modification, which in the course of ages chanced to arise, and which in any way favoured the individuals of any of the species, by better adapting them to their altered conditions, would tend to be preserved; and natural selection would thus have free scope for the work of improvement.[4]

Certainly among Darwin's key insights was that any "slight modification" favoring an individual's survival as conditions altered arose by chance.[5] Unequivocally, Darwin concluded, "Unless profitable variations do occur, natural selection can do nothing."[6] Changes in vegetation did not cause horses to birth colts with high-crowned teeth any more than tug-of-war with crocodiles lengthened elephant trunks. Mouth parts varied fortuitously, not on purpose, in ways that keep food coming.

Darwin concluded that a slight advantage—"still further modifications of the same kind"—might set in motion a positive feedback loop:

> Extremely slight modifications in the structure or habits of one inhabitant would often give it an advantage over others; and still further modifications of the same kind would often still further increase the advantage.[7]

The story of the evolution of horse dentition parallels the story of the elephant's trunk, the giraffe's tongue, and the rhino's lips. These traits extended the reach of their mouths, each "further modification of the same kind" increasing the advantage.

Phytoliths and Photosynthesis

The study of John Day horse fossils reaches completion with a unit titled, "You Are What You Eat." Its lessons link the local, particular case of the evolution of molar teeth among horses to Charles Darwin's seminal idea, natural selection.

Students enter real-world data provided by the teacher on a chart of "Molar Length for North American Horse Fossils" (horizontal axis) with respect to time during the past twenty million years (vertical axis). The span of time on the Molar Length chart extends the pattern entered on the Body Length chart through the early Miocene to the present. The traits scrutinized switch

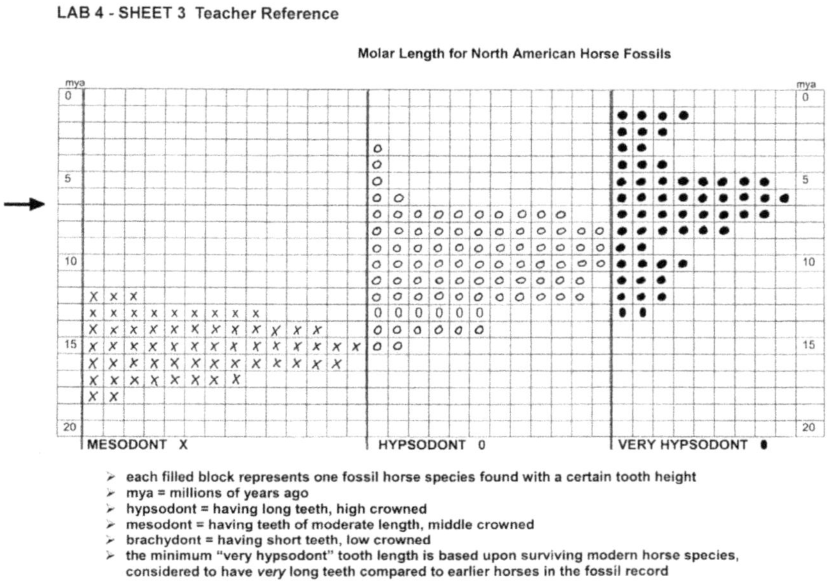

Figure 12.3 The Completed Graph of Increase in Molar Teeth Length Among Members of the Horse Family in North America During the Past Twenty Million Years from Lab 4 in the Horse Fossil Study Kit. Each filled block represents one fossil horse species found with a certain tooth height. *Source*: Graphic by John Fiedor, Chief of Interpretation at John Day Fossil Beds. Reproduced by permission of John Day Fossil Beds National Monument.

from body length to tooth type, but the trend remains the same. *Equus* is the champion of high-crowned molars and its very hypsodont teeth, for now, are the end point of horse evolution (figure 12.3).

Larger horses adapted to their own "prey," grass, with cementin-hardened, high-crowned (hypsodont) teeth. Grass fought back with weapons fashioned from silica known as "phytoliths" (plant-stones) that wore down teeth. The size and position of skull sockets recorded this arms race as space for bigger and better (at least in the context of munching on silica-enriched blades of grass) premolars and molars expanded.

"You Are What You Eat" stresses that horses are made from grass, just as grasses are made from air, primarily the carbon dioxide component. Chains of carbon atoms dominate plant molecules.

Grasses vary in their manner of linking carbon atoms to become sugars. The metabolism of three-carbon chain production (C3) grasses favors cooler, temperate climates; warm, dry tropical savannahs offer better opportunities for their four-carbon fixing (C4) siblings.

When C4 plants synthesize sugars, they utilize a higher proportion of atmospheric carbon 13 than do C3 plants. By the time carbon ends up in horse teeth, the sugars have long been digested, but the carbon remains. Deciphering the chemical composition reveals their dieting secrets. The ratio of carbon 13 to carbon 12 in horse teeth depends upon the proportion of C4 versus C3 in the plants they have consumed. Thus, a carbon ratio set by plants embeds itself throughout the body of the horse.

The body rots away, but fossilized teeth may endure for millions of years, preserving the carbon isotope signature of their owner's feeding habits. The chemistry of fossil teeth from a horse that grazed in a warm, dry grassland differs from that of one that lived in cooler, temperate lands. Carbon 13 enrichment indicates greater grazing on C4 plants.

When Charles Darwin dug a fossil of a horse tooth out of a cliff in Argentina, he probably gave little thought to its chemistry. Modern climate scientists, however, treat fossil horse teeth (and skeletal bones) as Pleistocene thermometers, though variation in latitude, long migrations, and dietary preference for C3 or C4 plants can skew the carbon ratio and complicate the picture.[8]

The teacher stands before the class with a horse fossil tooth in hand and says, "You are looking at a thermometer." The follow up question is, "How do you suppose it works?" To understand the method requires new knowledge in answer to the Horse Fossil Kit's essential question, "What does the geology and fossil record of the John Day Fossil Beds reveal about the climates and ecosystems of eastern Oregon's past?" Quite a bit, and that's its value![9]

NOTES

1. International Museum of the Horse, *Hyracotherim* (*Eohippus*), accessed June 10, 2020, http://imh.org/exhibits/online/legacy-of-the-horse/hyracotherium-eohippus/. The name "Eohippus" means "dawn horse."

2. *Horse Fossil Study Kit, Lab 3: Making a Horse Family Tree* (Kimberly, OR: John Day Fossil Beds National Monument, 2006).

3. Christine Janis, "An Evolutionary History of Browsing and Grazing Ungulates," in *The Ecology of Browsing and Grazing*, eds. Iain J. Gordon and Herbert H. T. Prins (Berlin: Springer-Verlag, 2010), 21.

4. Charles R. Darwin, *On the Origin of Species by Means of Natural Selection*, 6th edition (London: Murray, 1872), 63–64, ed. John van Wyhe, *The Complete Work of Charles Darwin Online* (2002), http://darwin-online.org.uk.

5. Curtis Johnson, *Darwin's Dice* (Oxford: Oxford University Press, 2015), xxiii.

6. Charles R. Darwin, *On the Origin of Species by Means of Natural Selection, Or the Preservation of Favoured Races in the Struggle for Life* (London: Murray, 1859), 82, ed. John van Wyhe, *The Complete Work of Charles Darwin Online* (2002), http://darwin-online.org.uk.

7. Ibid.

8. José L. Prado, Begoña Sánchez, and María T. Alberdi, "Ancient Feeding Ecology Inferred From Stable Isotopic Evidence from Fossil Horses in South America over the Past 3 Ma," *BMC Ecology* 11, no. 15 (June 14, 2011), https://doi.org/10.1186/1472-6785-11-15.

9. This chapter synthesizes anecdotes and observations spanning several years of my teaching as well as discussions with the John Day Fossil Beds National Monument staff about the development and use of the Horse Fossil Study Kit. I have written the story as a continuation of Ms. B.'s lessons on evolution. Chapter 11, "Skull Sockets," accurately reports activities and quotes from her classroom; Chapter 12, "Crowned Molars," idealizes instruction as outlined in the Study Kit. Ms. B. did enroll in the field study course described in the next chapter, "Hell's Pig."

Chapter 13

Hell's Pig

Ms. B. and her fellow teachers rested in the shade of a grove of apricot trees near the headquarters building of the John Day Fossil Beds National Monument. Heat had ended a blistering August morning of fieldwork in Blue Basin with the monument's paleontologist, Ted Fremd.

Fremd had demonstrated how to excavate a fossil by encasing it in burlap and plaster. The method has worked for over 100 years. Prior to the demonstration, Fremd assigned pairs of teachers to scout a bare slope of shale and clay for white shards—possible fossil fragments exposed by recent rains—and mark their location with small flags. Flagging had the potential of locating the source bed. In a previous summer, Fremd had unearthed a saber-toothed cat at a nearby location in this manner.

At the end of the exercise, the class decided to take one more hike along the upper trail that circumscribed Blue Basin. Midway through this hike, a flash of white against the bluish shale off the trail caught one of the teacher's attention. Bone? Yes, indeed—and not just a fragment. The bone turned out to be the fibula (the thin bone on the outside of the lower hind leg) of an extinct rhinoceros.

Refreshed by their shady respite, the class adjourned to the paleontology laboratory where Fremd's colleague, Connie Evans, was working to "liberate" the jaw of an oreodont (*Eporeodon occidentalis*) recently discovered near the monument's most colorful badlands section, the Painted Hills. For twenty-eight million years the animal's mineralizing remains lay buried in the greenish-blue rocks of the Turtle Cove Member of the John Day Formation before being plaster-jacketed away and picked at with dental tools.

TEETH CROWNS

What's an "oreodont," other than a once common, now extinct creature whose name rhymes with "hypsodont"? Oreodonts, ruminant mammals with paws (not hooves) and long tails, were more closely related to even-toed pigs, camels, and cows than to odd-toed horses and rhinos ("odd" and "even" refer to number).[1] Maybe they looked a bit like shorn sheep.

Given their low-crowned molar teeth, oreodonts likely browsed, but not on cookies. Their name has ancient roots: "oreo" stems from a Greek word for "mountain," and "dont" from the Latin for "tooth" (the English "dent" evolved from this common linguistic ancestor as well). Few scientists refer to these creatures as "cookie-tooths." If they did, school science would be more satisfying, as are Oreos. Perhaps there roamed among the Oligocene oreodonts a population with perfectly aligned teeth: the "orthodonts."

Playfulness with "dont" talk brings to attention a categorization scheme useful for distinguishing among teeth: brachydont, mesodont, hypsodont, and very hypsodont molars. These terms mean low-crowned cheek teeth (as found in humans), medium-crowned (deer), high-crowned (cattle), and *really* high-crowned teeth (modern horse, zebra, burro), respectively.

Evans moved on to a stunning fossil of a rhinoceros skull (*Diceratherium annectens*) entombed in grey, ashfall tuff. Her drill buzzed as she delicately chipped away and shared, "It's such a complete skull. It's a beautiful skull. Usually, this type of rhino has lost this nasal portion. It's very thin, very fragile—but we have it all here."[2] Grazing rhinos share a common ancestor with horses and depend upon hypsodont teeth.

Meanwhile, Ted pulled from his pocket another exquisitely preserved, precious find, a mountain beaver molar (family *Myagaulidae*, an extinct group and offshoot of the seldomly seen, modern mountain beaver). Often small animals do not range widely from their homes (rats, an exception, do cruise oceans). Teeth tell of diet, diet of local resources, and local resources of climate and habitat. Megabeast chompers demand attention, but minibeast teeth pack information, too.

PAINTED HILLS

Turtle Cove rock outcrops at the Painted Hills and elsewhere in the John Day basin. Once volcanic ash, the shales of the Turtle Cove weather into clayey soil. Rainfall erodes the exposed clays, turning the ground into an exhumed, sticky graveyard. Skeletal remains and scattered teeth, a textbook of time, make their appearances. At every scale beauty abounds. Look closely and the rocks reveal themselves as an assemblage of tinted fossil soils, ancient, weathered, ash lithified and re-weathered. Tiny pillars of calcite trace where roots once grew. Few plants find home on the popcorn-textured clay surface: when rain falls, the montmorillonite clays expand making germination difficult.[3] Stark, bare, and multicolored, the clays paint the landscape and, at flood stage, intermittent streams laden with sediment run striped in red and green.

Dark spheres and elongate, bulbous shapes, the size of marbles, catch the eye. Pebbles, perhaps? Congealed droplets of lava splattered on ancient ground? No, they are the concretized traces of sweat bees.[4] Deeper traces of burrows branch to end at rounded molds, the pupal chambers of dung beetles (*Pallichnus dakotensis*) that feasted on oreodont scat, horse manure, and rhino dung during Oligocene time.[5]

TRUNK FOSSILS

With an advance organizer in mind for examining fossil teeth, the class adjourned to the back porch of the monument's Cant Ranch historic home and museum. The park naturalist hauled out a heavy trunk, the John Day National Monument Horse Fossil Study Kit that afterward appeared in Ms. B.'s biology class at Willamette City High School.

The volcanic soils and strata of the John Day Fossil Beds, when read with expertise, vividly depict a story of Cenozoic mammals adapting to climate change. They persisted through times of spectacular tectonic events that wiped out populations with random abandon. Saber-toothed nimravids, cookie-toothed oreodonts, bulky brontotheres, massive chalcotheres, and equids made appearances. Dayville High School students captured parts of this story in their clay-action film, *Impressions of the Past*, viewed by Ms. B.'s biology students.

One of the teachers in the summer field class knew the region well. His great aunt had lived on the ranch adjacent to the monument's Sheep Rock Unit. In fact, her maiden name, Mascall, held a place of honor in the story of John Day basin fossils, for upon the John Day Formation rest the sixteen- to thirteen-million-year-old rocks of the Mascall Formation. These strata, named for the ranch where they prominently outcrop (along with ridiculously extensive beds of large, black cobbles), have yielded fossils of *Tephrocyon*, the "bear-dog"; *Dromomeryx*, the "giraffe-deer"; *Gomphoterium*, the four-tusked elephant cousin; and *Mylagaulus*, the horned rodent. All in all, a rather curious set of Mascall rascals. Trotting about among the weirdos were three different three-toed horses: *Merychippus*, *Parahippus*, and *Archaeohippus*.[6]

A single-family trunk of fossils seems inadequate for representing the travel through time of John Day's mammals. Horses, however, do one of the best jobs of leaving behind traces of their transitions. The Mascall's *Merychippus*, for example, marked the entry of high-crowning teeth useful for grass-feeding.

Field study immersed Ms. B. and her fellow teachers in the aesthetic grandeur of the badlands, basins, mountains, and canyons of John Day Country and its fossil cornucopia. Stopping at the monument's interpretive center introduced them to the Horse Fossil Study Kit. The experience enabled them to teach science grounded in local story, to link backyard discoveries to a much bigger picture, and to incorporate the aesthetics of museum quality artifacts into learning about evolution.

TERMINATOR PIGS FROM HELL

The ruins of a little schoolhouse, unoccupied since 1936, once graced the entrance to "The Cove," a rimrocked, open expanse of rangeland, drained by the intermittent Pine Creek. Disassembled, moved, reconstructed, and fully restored, the schoolhouse—free of cow pies—survives as an interpretive center in Fossil, Oregon.

The school stood where once roamed the Hell Pig, an extinct entelodont with ancestral affinities to "whippos"—the common ancestors of whales and hippos. Entelodonts, bulldozer-sized, tusked beasts with massive heads and shoulders, terrorized their neighbors for more than twenty million years. In

Figure 13.1 Skull of *Archaeotherium Mortoni*, an Entelodont Popularly Known as the Hell or Terminator Pig from the Oligocene Epoch. *Source*: Photograph by James St. John of a specimen exhibited at the Carnegie Museum of Natural History. Creative Commons License Attribution 2.0 (CC BY 2.0).

Miocene time, as recorded in Mascall rocks, the last of the Hell Pigs met its extinction fate.

The Ratfish Wranglers, a group of musical paleontologists from Alaska, immortalized the Hell Pig in a lovely ballad with lyrics describing it "gorging on his putrid feast of flesh . . . standing like the Devil's proof of Hell."[7] Ted Fremd refers to this creature, whose closest living relative is the hippopotamus, as the "Terminator Pig." Apparently, the beast, by either name, was a badass, badlands creature (and ballad) for the ages. Its skull (figure 13.1[8]) resembles the heads of the dragons in *Game of Thrones*.

Noted for archaeological treasures and the occasional fossil tooth of the Oligocene age North American entelodont, *Archaeotherium caninus*, the Cove's amphitheater provided the perfect ambiance for geologic field study. Ms. B. and her classmates spent a grueling day there mapping paleosols, cliffs of volcanic tuff, lava flows, lithified lake bed sediments, ridges of ignimbrite, and weathered paleosols. They arranged these strata as series in time, correlated with the fossils they had learned about at the monument.

FOSSIL TOWN

At the end of field study week, the class journeyed to the Fossil, Oregon, passing the spruced-up Pine Creek schoolhouse as they drove into town.

Down the street, past Fossil Mercantile, they could see a sign in front of a church that read "Fossil Bible Study, Wednesday Evenings." The van pulled into the parking lot of the local high school. Behind its athletic fields, a pile of broken rock filled a small trench excavated by a Bridge Creek flora researcher with a backhoe. Above, the trench fragmented rock tumbled down a steep slope. Erosion, abetted by humans with picks and shovels, further exposed an Oligocene lake bottom. Turned into layers of stone, its seasonal deposits of sediment had trapped the leaves of ancient autumns.

Thirty minutes later, having paid a small collection fee to the high school, everyone in the class carried away in their daypacks fossil impressions of *Metasequoia* (dawn Redwood), alder leaves, and other plant fossils evocative of a Tennessee woodland, not Fossil's bone-dry hills.

A MOST PLEASANT PLESIOSAUR

They reluctantly departed the high school's popular excavation site to arrive on time for a scheduled visit to the spacious and air-conditioned Oregon Paleolands Center, opened in 2009. The center features exquisite samples of the fossil flora from the Fossil township and its John Day environs together with examples of mammal quarry discoveries. The teachers spent several minutes matching their finds to the museum specimens. "You can call me Mitch," barked a hoarse voice, seemingly from nowhere.

One of the teachers, an ersatz ventriloquist, stood next to a creation of paleoart labeled, "The Mitchell Plesiosaur." Its mouth agape and dagger-teeth threatening, Plesiosaur Mitch growled out a few more syllables, "Come closer, precious ones, the better to teach you."

Part fossil bone, part iron rebar, and welded all over, the Mitchell Plesiosaur's presence mesmerizers onlookers, as did the first plesiosaur fossil, discovered by England's famed fossil hunter, Mary Anning, in 1823.[9]

Unearthed in 2003, the Mitchell skeleton became a sensation in thinly populated Wheeler County—a marvelous monster with a paddle-limbed body behind a long, skinny neck ending with a head packed with reptilian teeth. The high school vocational education class took on reconstructing the twelve-foot-long creature for display. Noted paleo-artist (and welder) Larry Williams worked with Mitchell High School's vocational education students, and elementary school children composed a book telling the plesiosaur's story.

Geologist, author, and photographer Ellen Morris Bishop made an appearance and described plesiosaurs to the class as "the tigers of the Cretaceous seas" eighty million years ago.

The Mitchell Plesiosaur paddled the ocean. Then its world changed. A line of volcanoes formed to the east, parallel to a subduction zone as the margin of North America and the floor of the Pacific Ocean engaged in geologic combat. The violence subsided, then reignited. As a succession of collisional fronts migrated west during Cenozoic time, North America added terrestrial habitat. Deeply agitated, the mantle of the earth roared its presence, erupting marine volcanoes in the style of Mauna Loa, now docked as mountains interspersed in Oregon's Coast Range.

Quenched for a time beneath North America, the mantle plume probed and poked from below, looking for a zone of weakness. Slabs of North America's foundation blocked its passage while the continent itself continued its leisurely drift westward. At just over twenty-nine million years ago, near today's Prineville, Oregon, when early Oligocene's *Mesohippus* and *Miohippus* were tiptoeing about, magma scorched its way to daylight. Eruptions shattered the crust, launched shards of rock, billowed clouds of glowing hot ash, and spewed streams of molten lava.

The detritus of Hades rained down to smother the forested land, cremating oreodonts, rhinoceroses, horses, and sabertooths. Lava domes, emptied of molten rock, collapsed. Ashfall tuffs filled and frozen lavas ringed the sunken calderas formed by Oregon's "supervolcano" (a "caldera" is the structure left behind when an emptied magma chamber collapses).[10]

Near the modern Crooked River to the east of Fossil, volcanic collapse left rugged cliffs, beloved by the climbers who frequent Smith Rock State Park. These volcanic events probably resembled the processes that created the Yellowstone Caldera. Radiometric dates from numerous ash beds and lava flows from Crooked River region eruptions number the pages of the John Day fossil storybook. Yellowstone National Park may well be the present location of the plume that created Prineville's Crooked River Caldera. It's a bit scary to think it still exists.

Mitchell's environs host the signatures of terrestrial catastrophes. Its quiescent volcanism has buried an uplifted world that was marine fifty million years ago. In Cretaceous time, the sands, pebbles, silts, and clays, of a near-shore environment, the eighty-million-year-old rocks of the Hudspeth and

Gable Creek Formations, accumulated on ocean floor destined to become near desert.

The Mitchell Plesiosaur swam and died in these waters while the west coast of North America lay to the east in today's Idaho. In time's vastness, the earth twisted, folded, and faulted just right, juxtaposing the plesiosaur's graveyard with the volcanic tombs of terrestrial beasts. In 1899 a team excavating cliffs near Mitchell above Bridge Creek discovered an exquisitely preserved skull of an *Entelodon*.[11]

The Terminator Pig from Hell and the Mitchell Plesiosaur, local residents separated in time by tens of millions of years, have become museum neighbors telling a story of creation, of destruction, of wonder, of evolution. They represented monsters from paleontology's id tamed as memories of landscapes long vanished.

Refreshments were served to the visitors at the Paleolands Center. Mitch, the oldest fossil on display, rested on his perch, contemplating the chapters that had unfolded since the demise of his cherished Cretaceous seas. The Pine Creek schoolhouse stood silently a few blocks away, it too rescued from the sands of time. And all around in the rocks of ages more teeth and skulls awaited discovery by lucky fossil prospectors.

NOTES

1. Gregory J. Retallack, Erick A Bestland, and Theodore J. Fremd, "Reconstructions of Eocene and Oligocene Plants and Animals of Central Oregon," *Oregon Geology* 58, no., 3 (May 1996): 51–69.

2. *Oregon Field Guide*, season 5, episode 507, "World Class Fossils," produced and narrated by Steve Amen, aired on February 2, 1994, on Oregon Public Broadcasting.

3. Stuart Garrett and Berta Youtie, "Oregon Plants, Oregon Places: The Painted Hills: Thirty Million Years of Phytography," *Kalmiopsis* 2 (1992): 21–31.

4. Gregory J. Retallack, "Trace Fossils of Burrowing Beetles and Bees in an Oligocene Paleosol, Badlands National Park, South Dakota," *Journal of Paleontology* 58, no. 2, (March 1984): 571–592, accessed May 22, 2020, https://www.jstor.org/stable/1304803.

5. Gregory J. Retallack, "Late Oligocene Bunch Grassland and Early Miocene Sod Grassland Paleosols from Central Oregon, USA," *Palaeo: Palaeogeography, Palaeoclimatology, Palaeoecology* 207 (2004): 203–237.

6. National Park Service, "Mascall: John Day Fossil Beds National Monument." This site includes a photo gallery of the assemblage of creatures discovered in the Mascall Formation, accessed September 9, 2020, https://www.nps.gov/joda/learn/nature/mascall.htm.

7. Ray Troll, Russell Wodehouse, and the Ratfish Wranglers, "Hell Pig Ballad," *Fish Worship*, compact disc, track 12, recorded at the Chimera Club, Ketchikan, AK (2015).

8. James St. John, *Archaeotherium mortoni*, fossil giant pig skull (Middle Oligocene, Colorado), accessed September 9, 2020, https://www.flickr.com/photos/47445767@N05/31459808453. Creative Commons License Attribution 2.0 (CC BY 2.0), https://creativecommons.org/licenses/by/2.0/.

9. Two books for young readers share the same title in telling stories about significant contributions to paleontology by women: (a) Janet Ivey-Deunsing, *Daring to Dig: Mary Anning: Fossil Hunter*, illustrated by Katie Grayson (Franklin, TN: Traitmarker Books, 2020), and (b) Beth Stricker, *Daring to Dig: Adventures of Women in American Paleontology*, illustrated by Alana McGillis (Ithaca, NY: Paleontological Research Institution/Museum of the Earth, 2017). The Paleontological Research Institution also sponsors a traveling exhibit, *Daring to Dig: Women in American Paleontology*, accessed June 10, 2020, http://www.daringtodig.com/.

10. Jason D. McClaughry, Mark L. Ferns, Caroline L. Gordon, and Karyn A. Partridge, "Field Trip Guide to the Oligocene Crooked River Caldera: Central Oregon's Supervolcano, Crook, Deschutes, and Jefferson Counties, Oregon," *Oregon Geology* 69, no. 1 (2009): 25–44.

11. *Terminator Pigs of Eastern Oregon*, "Terminator Pigs" by Ted Fremd (May 15, 1999), and "Entelodon Legacy," by Tonia Seebart (July 13, 1998). John Day Fossil Beds National Monument, accessed June 10, 2020, https://www.nps.gov/joda/learn/nature/terminator-pigs.htm.

Chapter 14

Vivid Canyons

GEOLOGIC LANDSCAPES

Geoscientists from an array of disciplines—paleontologists, volcanologists, climatologists, and more—value landscape art.[1] Artists accompanied early scientific and mapping expeditions of the American West. For his second journey down the Colorado River in 1871, John Wesley Powell, for example, recruited Thomas Moran as a "guest artist" whose masterpiece *The Chasm of the Colorado* resides at the Smithsonian American Art Museum. Artists' and geologists' perceptions of patterns in the structure of the earth complement each other.

Metaphor

Metaphors are ubiquitous in geologic descriptions—tectonic "plates" is the most obvious example. Throughout a distinguished career in geology and science education, William D. Romey promoted the incorporation of art in teaching the geosciences. Romey underscored how recognizing the "metamorphic nature" of terminology toppled preconceptions and "entrapment by ruling paradigms."[2] Metaphors can leverage conceptual change.

To interpret a landscape, whether a Pacific lagoon or an Argentine river terrace, some features are neglected to discover truths about others. With time, neglected elements, based upon new conceptualization, may lead to interpretations that displace those trusted previously. In John Day Country, ashfall tuft and tephra, once thought to have spread from a chain of volcanoes, turned out to be the crater rim deposits of a vast caldera. The framing changed, but the truths of habitat succession, speciation, and extinction marched forward.

When testing a hypothesis, scientists must decide what observations merit effortful attention and what features may be productively ignored. The design of an experiment imposes, as does art, boundaries, often separating the object of interest from the host of its attachments to the natural world. Examining the few that matter most with a particular purpose in mind means making constructive judgments about what to neglect.

Dance

While a faculty member of Indiana University's School of Art, Architecture + Design, Bonnie Sklarski prepared her mind and senses to paint landscapes by studying structural geology. Instead of seeing a static expanse of multicolored rock, Sklarski witnessed the movements of the earth, events from one process disrupted by another. Having painted landscapes since childhood, Sklarski found them to be the opposite of still life paintings

In her 1984 *Man in the Wilderness*, five figures scamper among rocks of a creek set in a deep gorge. Close inspection finds vibrant life, assertive yet diminutive. Having heard the clink of footsteps on rock, an antelope turns its head. The cliffs tower above in a commanding, yet grudgingly beneficent, repose. Rock seems to remind the human figures of its harsh indifference to their escapades. Boulders tumble from above to leave obvious gaps in the rimrock. The landforms follow their own paths through time.

The landscape does not keep still. Shadows fade; wind rustles. Structures shift to rhythms difficult for humans to perceive. "A change in color of a rock means more than putting away one tube of paint and bringing out another. It celebrates that ancient season when drying foliage accumulated and was layered over a fine silt at the bottom of some ancient pond."[3] The scene requires choreography.

Sklarski's paintings express, in her words, the "spellbinding mystery, power, and sense of the universe which the infinite variety of nature and the monumental forces of geologic movement hold over the human imagination."[4] With a few figures and some rocky cliffs in an imagined place, Sklarski located in the particular what she believed to be universal.

Her imaginary landscape has the feel of Dry Canyon, a place not far from the John Day Fossil Beds and the Thomas Condon Paleontology Center. The

Center's horse fossil curriculum defines cognitive objectives aligned with state standards and sets as its aim sharing the experience of paleontological inquiry. To achieve this aim aesthetic expressions embedded in student activity, not just aesthetically crafted instructional materials, are essential. The curriculum's collection of fossil teeth and their corresponding skulls choreographs a spellbinding mystery where monumental forces of geologic movement dance with the infinite variety of nature.

The students of Dayville High School worked with professional artists to produce a film, *Impressions from the Past*, both scientifically valid and whimsically entertaining. No doubt they enjoyed the task of sculpting prehistoric, stop-action figures turning their heads as volcanoes erupted. The film has aged well; it still captures interest.

SKETCHY ZOOS

Study of the Horse Fossil Kit calls for interpreting diagrams, constructing graphs, and measuring artifacts—visual tasks. Challenging students to draw animal forms develops faculties for visualizing evolution further. A zoo is a good place to begin. No exhibit is without art, and zoos pay close attention to principles of artistry in crafting visitor encounters with animals. A charcoal sketch portrays a chimp's pensive facial expression; a bronze sculpture freezes a polar bear's efficient paddle; a series of panels isolates how a penguin displays aggressive and submissive behavior. Exhibits necessarily neglect drawing attention to some features of complex behaviors in order to emphasize others.

Stick Figures

When learning to sketch at a zoo, achieving realism, elegance, and complexity is not the goal. The line is the thing. Line segments depict angles and lengths. From shoulder to toe the segments zig then zag, some short, some long. The back may arch, run straight, or sag. A thumb moves along a pencil held in the line of sight, the pencil's angle and the segments marked by the thumb's position transfer to paper in correct proportion and orientation.[5]

Each change in angle implies a joint, shoulder to foot. The sketching mantra does the trick: make a stick figure, block in basic shapes, contour the edges, add some shading. Finish with details to taste. Work proceeds quickly at each stage.

Illustrations of stick rhinos and stick hippos appear. The elephant's massive body reduces to skinny lines and angles. The necks of storks and giraffes seem almost silly in proportion to their bodies. Trapezoids, triangles, squares, circles, ovals, and rectangles create heads, necks, limbs, haunches, shoulders, ears, muzzles, hooves, and tails. Soon a stocky mountain goat, then a bulbous warthog, later a rotund orangutan, and eventually a long-limbed gibbon peer back from the pages of the sketchbook.

Revisiting Comstock

Fine-tune the digits. While looking, think of questions as taught by Anna Botsford Comstock of Nature Study fame. Do claws, hooves, or nails come in singles, doubles, triples, or more? Do they turn inward or outward or backward or align straight ahead? Which way do the hind toes point on creatures that climb: the sun bear, the margay cat, the local squirrel that chanced to run by? Note which parts of the foot or hand touch the ground when the creature walks and whether these contact points are the same or different when its stands or runs.

Two steps remain: (a) contouring the outline of the animal's profile and (b) finishing with shading. The process is Picasso in reverse: contouring the blocky impressions yields more realistic forms. Adding details of shading plus impressions of feathers or fur completes the entry.[6] Sketch artists are now ready to take on the task of expressing the features of creatures from John Day Country as perceived in their imagination. The hand trains the eye, and the eye stimulates the mind.[7]

CRITERIA OF ARTISTRY

Works of art, in the judgment of the noted art educator, Elliot Eisner:

1. Make the obscure vivid and empathy possible;
2. Direct attention to individuality and locate in the particular what is general;

3. Possess a sense of wholeness, an organic unity necessary for aesthetic experience and credibility;
4. Neglect some features of the subject in order to discover truths about others.

In an "artistically crafted work" the particular bears witness to the universal.[8] The same can be said of thinking Darwinian or other disciplined thoughts. A scientific theory connects aspects of the world to unveil a pattern of meaning capable of explaining natural phenomena. Ideas cohere in theory, depending upon each other to derive their respective meanings as they map onto features of nature. Without coherence, credibility fails. "Coherence, imagery, and particularity are the fruits of artistic thinking" argues Eisner.[9] Coherence, imagery, and particularity are equally the hallmarks of disciplined thinking and the attributes of an excellent curriculum.

Art frames a subject and distills perception to yield a meaningful experience, as should science teaching. Similarly, the neglect of some truths to discover others is an essential component of inquiry. Scientists must decide what to neglect when designing an experiment, their conceptions functioning to frame an inquiry. Forms chosen to represent the data (table, tree, map, or equation) capture features of reality from one perspective while omitting others, as do various traditions of art.

The coherence of a scientific theory connects observations of the particular to wider contexts. Scientific facts, analogous to works of art, "make the obscure vivid." In science, as in art, the starting and end points are empathy for the subject, a grasp of its meaning, and a feeling for its significance.

Eisner's criteria of artistry work equally well as superordinate aims for the practices of science and curricula of schools. Eisner's criteria remind science educators of the importance of emotional content to learning, an integration that prevails in natural history.

Empathy

No natural history artist has surpassed the skill of John James Audubon. His *Birds of America* inspired generations of American naturalists and conservationists. From 1827 to 1839 he worked with an engraver to prepare 435 images of birds on copper plates from his original watercolor paintings. Their dimensions, slightly more than two by three feet (for printing on "Double Elephant" paper in the jargon of the age), accommodated life-sized drawings. The engravings were hand colored.[10]

Audubon's *Birds of America* has intrinsic value as objects of art and instrumental value to society as inspiration for conserving nature. He helped others see what he saw—and why he cared. Audubon's paintings illustrate the meaning of Elliot Eisner's "criteria of artistry," the first of which is "vivid depiction." Audubon's prints are vivid and valued. They create empathy for his subjects.

Bound sets of seven volumes included Audubon's field notes for each painting.[11] Today, prices range widely. At auction in 2013 one of the one hundred and twenty complete original sets sold for eleven and a half million dollars.[12] In order to make the collection more accessible, Audubon crafted much smaller images (approximately ten by six inches) from camera lucida projections.[13] He printed this set using lithographic limestone, and studios produced Audubon's lithographs until 1871.

"What a treat for me to disclose things unknown to all the world until now!" wrote Audubon in a letter to his close friend, the Reverend John Bachman, in 1835. Audubon's vivid depictions of birds disclosed "things unknown" in natural history and his particular works have held universal acclaim among conservationists.[14]

Paradox

From Eisner's perspective, vivid depiction makes empathy possible. Aesthetic insight enables empathetic imagination, and empathy is an ideal guide to action. Vivid depiction comes first. A work of art has individuality. Despite its singular nature, aesthetic expression works best when it prompts attention to find what is universal located in the particular. From a small bone in the wrist of a fossil grew the story of avian descent from bipedal dinosaurs—and then the Jurassic Park movie franchise.

Many disciplines value the qualities Eisner finds in art: how aesthetic expression distills perception to yield a particular experience. To make inquiry feasible, the scientific investigator must exercise, as does the artist, "constructive neglect."[15]

The plastic replicas of fossil horse skull and teeth discovered at the John Day Fossil Beds National Monument illuminated an exhilarating slice of reality. The art of the artifact attracted attention and piqued interest; the fossils held climate change clues. Teeth spoke of ecological conditions, the skulls of ancestry. Skull and limb anatomy indicated the degree of common ancestry, a judgment deeming some observations as more reliable as markers of inheritance than others.

Properly framed, ancient remains reveal stunning stories. Art elevates some details and ignores others according to what the artist feels matters. The artist's craft is to "create a paradox," claims Eisner. The paradox is to reveal "what is universal by examining what is particular,"[16] to find the story of creation etched in a fragment of rock.

NOTES

1. William Dowden Romey, "Using Patterns, Icons, Abstractions, and Metaphors from Art in Geoscience Classes," *Journal of Geoscience Education* 48 (2000): 295.
2. Ibid., 354.
3. Susan Wintsch, "Man and Nature," interview with Bonnie Sklarski, *Research & Creative Activity*, Indiana University Bloomington, 8, no. 1 (September 1984): 9.
4. Ibid., 11.
5. Cathy Johnson, *The Sierra Club Guide to Sketching in Nature* (San Francisco, CA: Sierra Club Books, 1997), 37.
6. Claire L. Walker and Charles E. Roth, *Keeping a Nature Journal: Discover a Whole New Way of Seeing the World Around You* (North Adams, MA: Storey Publishing, 2003), 111, 181–185.
7. The section "Sketchy Animals" is abridged from Charles R. Ault Jr., "Out on a Limb: Sketching at the Zoo to Compare Limb Anatomy," chap. 6 in *Do Elephants Have Knees? And Other Stories of Darwinian Origins* (Ithaca, NY: Comstock Associates, an imprint of the Cornell University Press, 2016), 97–101.
8. Elliot W. Eisner, *The Kind of Schools We Need* (Portsmouth. NH: Heineman, a division of Reed Elsevier, 1998), 152.
9. Ibid., 154.
10. "Audubon Prints," Audubon House Gallery of Natural History, accessed June 27, 2020, http://audubonhouse.org/audubon-prints.
11. Ibid.
12. Mike VanHelder, "The World's Most Expensive Birdbooks: John James Audubon's 'Birds of America,'" *Audubon News* (May 2, 2016), https://www.audubon.org/news/the-worlds-most-expensive-bird-books.

13. A camera lucida's normal mirror reflects an image of the object of interest onto a semitransparent mirror. The artist looks through the semitransparent mirror at the drawing surface. The image, its size dependent on the distance from the camera lucida to the object, appears to be on the drawing surface where the artist may trace it.

14. George Dock, Jr., "Audubon in the Atomic Age," *The Audubon Folio* (Harry N. Abrams, 1964). This edition contains forty Audubon prints and Dock's biographical sketch of Audubon.

15. Ibid., 153.

16. Eisner, *The Kind of Schools We Need*, 152.

Chapter 15

Fossil Imagery

Children delight in making prints from a potato sliced in half. Here's how it works: cut away everything except the lines and shapes to print. The image carved into the potato appears in relief. Dip in tempera paint or ink and press or stamp on paper, cloth, or rock. Voilà! Repeat to make many more prints.

In learning this art form, primary grade and preschool age children participate in a centuries-old innovation crucial to the wide dissemination of scientific illustration.

ON THE ORIGIN OF PALEONTOLOGY

The birth of paleontology, in the opinion of historian Martin Rudwick, occurred on July 28, 1565. On that date Conrad Gesner completed *On Fossil Objects, Chiefly Stones and Gems, Their Shapes and Appearances*. The title suggested the significance of visualization—shapes, appearances—to the study of fossils. Prints from woodcuts had made the wide dissemination of scientific illustration possible.

To Gesner and his contemporaries, "Fossil" meant something quite different than to paleontologists today. The category included any naturally occurring object or material dug up from the earth or perhaps just lying about on its surface. There had to be some pattern that captured attention and begged for an explanation. Ores, gems, ammonites, mineralized bone, shells, and crystals fell under the umbrella of "fossils."[1]

Paleontology began during an epoch when science, the occult, and religion were poorly differentiated from each other. Calculating and conjuring, prayer and prophecy, magic and medicine, speculation and empirical observation swirled about in the minds of sixteenth-century scholars.[2] The shape of a plant could suggest its medicinal value. A fossil might form in rock according to the same process that yielded mineral crystals in geometrical shapes. Perhaps a bit of storm-tossed shell landed in mud where it seeded the growth of similar shelly forms.

The paleontologist's goal is to interpret the origins of objects from the past through comparison with those from both the present and the past. Progress in paleontology premised the notion that fossil's resemblance to living plants and animals was a clue about its origins. Fossil finds are first local, though some occurrences are widespread, and illustration, thanks to Gesner's woodcuts, became the means of sharing. It takes a community to interpret a fossil; accurate illustration binds the community of paleontologists together to achieve that purpose.

Gesner wished to communicate with precision the identity of a host of specimens long before the invention of photography or computer graphics. To compare, for instance, the fossil of a shell or a crab with its living counterpart he depended upon the technology of woodcut illustrations. Arguing similarity required accurate illustration. Illustrations mitigated "the hazards of inadequate means of verbal communication."[3] Gesner was certainly not alone in his use of woodcuts, but the number he used and the share of his message they carried represented a departure from contemporary norms. According to Rudwick, for this reason, his achievement marked the origin of paleontology as a disciplined field of study.

WOODCUTS AND COPPER PLATES

Descartes' seventeenth-century rationalism helped to disentangle scientific methodology from its occultish companions and elevate explanations based upon natural causes. Conrad Gesner may have felt comfortable in the company of sixteenth-century conjurers and alchemists. He published *On Fossil Objects* decades before the advent of Descartes' logic and rigorous skepticism. There are, however, elements of continuity in any science even across a paradigmatic chasm. Gesner advanced his science by better thinking through improved visualization, a means still pursued.

The Woodcut's Heritage

Japanese artisans were printing on textiles with woodcuts over 1200 years ago. By 1400, crude woodcutting had found its way to Europe, the prints benefiting by hand coloring. By 1500, woodcuts became a staple component of the new and consumer-friendly product, printed books. Albrecht Dürer's woodcuts, for example, the *Four Horsemen, from the Apocalypse* (1498), elevated the art to an unsurpassed level "with subtle gradations of tone and suggestions of texture."[4]

Woodcut techniques were well-developed by the time Gesner published *On Fossil Objects*. From art, a science was born. By 1830, publishers could print text and woodcut pictures inexpensively on the same pages, a boon for geologists publishing in the first half of the nineteenth century. They valued illustration as a means of explanation.

The Engraver's Art

Woodcuts worked well as line drawings and sketches accompanying text, but faced limits as precise, accurate, scientific illustrations. Cut too fine, wood easily broke. Once removed, slivers of wood could not be replaced to revise the drawing. The technology of copper engraving, known from the fifteenth century, overcame these limits. Woodcuts had to be large and carving wood, no matter the skill of the artist, restricted detail. Copper engraving, popular with mapmakers and in use since at least 1430, gave illustrators a boost. A copper plate could be used, altered, and reused. Engravings, however, were expensive commodities, still set troublesome limits to the resolution of detail, and required special paper for printing and binding.[5]

Illustrations printed from copper engravings often appeared at the end of book, a practice not unlike today's norm of downloading images linked to published manuscripts. Engravings improved the quality and precision of nonverbal communication in natural history, especially among paleontologists.[6]

The change from woodcuts to copper engravings was a significant advance in illustration. Switching from copper to steel engravings when printing thousands of copies reduced printing costs without sacrificing quality. By analogy, advances in instrumentation serve the same function in physics. Illustration in geology, like instrumentation in physics, increases the capacity for geologic inquiry to achieve its aims.

VIRTUAL FOSSILS

Authentic fossil objects or high-quality replicas (e.g., fossil skulls) in the hands of students increase their opportunities to engage in the practices of paleontology. Tangible artifacts prompt the deployment of techniques of investigation relevant in a particular context.[7] Keeping that relationship intact is a principle vital to good science teaching. In paleontology, attention to transitional fossils amplifies the importance of context, as exemplified by the John Day Monument's Horse Fossil Study Kit's focus on variation among specimens collected from the period of time when horses in North America transitioned from browsers to grazers.

A single horse tooth fossil, vividly prepared, whispers clues of diet, habitat, and ancestry. Tooth chemistry preserves the carbon isotope signatures of their owner's feeding preferences, and tooth morphology indicates, in general, a browsing or grazing lifestyle. The chemistry of fossil teeth from a horse that grazed in a warm, dry grassland differ from that of one that lived in cooler, temperate lands. Collections of specimens cohere as records of evolutionary adaptations to climate change. Carbon 13 enrichment indicates greater grazing on plants that photosynthesize chains of four-carbon sugars than on plants that cobble together three-carbon sugars.

High-quality images function in a similar manner similar to physical reproductions of actual artifacts. Collections of photographs of stunning fossils, for example, can be studied to examine the evolution of bird flight from dinosaurian ancestors without the expense of collecting feathered specimens in China's Liaoning Province or handling precious impressions of *Archaeopteryx*.[8] With guidance and superb images, novices can search for anatomical features in transition and draw conclusions about evolutionary relationships.

Schools, unless allied with a museum partner, often lack access to authentic specimens. The expense of travel restricts visiting places where the magic of holding a rare fossil of an extinct monster is commonplace. Casts and models, however, do make adequate substitutes—opening a trunk of equid or hominid skulls, even if plastic, spanning millions of years is a hypnotizing experience. But even these resources come with price tags for their use and curation. Fortunately, digital technology has a solution capable of expanding access exponentially.

The John Day Fossil Beds National Monument enthusiastically distributes its trunk of fossil horse skull replicas. The heft and look of the plastic models are as close as most students will ever be to holding the mineralized skull of a horse extinct for tens of millions of years. For the purpose of classroom engagement, however, reality does have a substitute—a direct descendant of the art of woodcuts, engravings, lithographs, scientific drawing, and photography. Its name is the Digital Atlas of Ancient Life.

By web-surfing to the Atlas, anyone, anywhere, can inspect a digitized, three-dimensional reproduction of a fossil from the Paleontological Research Institution's collection in Ithaca, New York. Just a couple of clicks sets up zooming-in to observe details or rotate each virtual specimen in simple wonder or as an object of research. The "enterprising investigator" is anyone interested in studying a replica of an oreodont fossil.[9] As other museums adopt this technology, schools will have free access to a stunning variety of ancient life-forms.

ART AS METHOD

The fossilized traces of the past display beauty and stand alone as art, as do the reconstructions of ancient ages in stippled drawings and bronze castings. The life of the past displayed in museum murals is the twenty-first century's equivalent of the paintings on the walls of the Lascaux Caves made 17,000 years ago. The art in both contexts evokes empathy for humanity's place in the midst of the mystery of life and amazement for the present moment. They create aesthetic experiences.

Paleontology cannot exist without illustration—an art form at the heart of methodology. How well can students learn evolution's bushy, branching story of John Day's three-toed horses, gracile rhinos, oreodonts, sabertooths, entelodonts, and their extinctions? How able are they to bear witness to the volcanic construction and destruction of landscapes of the past in the landforms of the present? To what extent do they understand the scale of today's ecological transformations by comparison to those of the past? Their original vivid, coherent, aesthetic expressions may tell.

Aesthetic expression, experience, and appreciation matter. They are not enhancements or enrichments. They are a means to teach science beyond standards. There is no separation of art from paleontological (and geological) history and the conduct of inquiry in these fields. The aesthetic dimension is central to interest, practice, and understanding—and even to evolution itself, both organic and cultural. Sensing beauty adds significance to learning, and the feeling of significance amplifies meaning. The aesthetic expression of thought deepens intelligence.

Aesthetic experiences created by expressions of scientific insight are open to everyone who takes time to look, study, and express their feelings, best done by crafting another artistic creation no matter how crude. Aesthetics is a matter of the heart, a feeling of having had a consummatory experience, of sensing something significant in the perception of an image or object, a sensation Conrad Gesner would appreciate.

NOTES

1. Martin J. S. Rudwick, *The Meaning of Fossils: Episodes in the History of Paleontology*, 2nd ed. (Chicago, IL: The University of Chicago Press, 1972).

2. Anthony C. Grayling, "Dr. Dee and the Potent Art," in *The Age of Genius: The 17th Century and the Birth of the Modern Mind* (London: Bloomsbury, 2016). Grayling describes how Shakespeare based Prospero, the protagonist in *The Tempest*, on a well-known scholar from that era: Dr. John Dee, court astrologist and adviser to Queen Elizabeth I. Dee studied mathematics and assembled an extensive library for seeking knowledge of nature's secrets presumed once known to ancient cultures. These secrets, he surmised, were being newly discovered through alchemy, the midwife of modern science. Dee was suspected of conjuring spells, perhaps the storm—that mangled the Spanish Armada and saved England from invasion.

3. Ibid., 6.

4. Wendy Thompson, "The Printed Image in the West: Woodcut," in *Heilbrunn Timeline of Art History* (New York: The Metropolitan Museum of Art, 2000–), accessed June 27, 2020, https://www.metmuseum.org/toah/hd/wdct/hd_wdct.htm.

5. Copper Development Association, Inc. "Copper Engraving Plates," accessed June 27, 2020, https://www.copper.org/education/history/60centuries/industrial_age/coppereng.html.

6. Rudwick, *Meaning of Fossils*, 7.

7. Eliza Estrup and Marianne Achiam, "The Potential of Palaeontology for Science Education," *Nordic Studies in Science Education* 15, no. 1 (January 2019): 97–108, accessed October 21, 2020, https://doi.org/10.5617/nordina.5253.

8. Marianne Achiam, Leonora Simony, and Bent Erik Kramer Lindow, "Objects Prompt Authentic Scientific Activities among Learners in a Museum Programme," *International Journal of Science Education* 38, no. 6 (2016): 1012–1035, accessed October 21, 2020, https://www.tandfonline.com/doi/abs/10.1080/09500693.2016.1178869.

9. Paleontological Research Institution, "Digital Library of Ancient Life," accessed October 5, 2020, https://www.digitalatlasofancientlife.org/vc/userguide/.

Chapter 16

Inspiring Invertebrates

A steel engraving, the frontispiece to Charles Lyell's first volume of *Principles of Geology* in 1830, holds special historical significance. The illustration impressed upon generations of geologists the value of interpreting the past through analogy to the present. Lyell's frontispiece defined the moment of birth of modern geology in the same sense that Gesner's woodcuts set the starting point of modern paleontology. Pictures framed their paradigms.

THE TEMPLE OF SERAPIS AT POZZUOLI

In Lyell's illustration, an observer looks across the Bay of Baiae in the Gulf of Naples at the pillars in the ruins of a second-century Roman marketplace, erroneously inferred in the caption as the "Temple of Serapis at Pozzuoli" (Serapis was the god of a Graeco-Egyptian cult initiated in the third century BCE). Three columns, forty-two feet tall, command attention.

Strangely, twelve feet above their pedestals the smooth marble changes in texture. Tiny openings lead to pear-shaped cavities several centimeters in depth, some with shells. These holes predominate for the next twelve feet. What created these ubiquitous chambers in the marble? Were they present when the columns were quarried?

Bivalve Dormitories

The subtitle of Lyell's book clearly stated the principle that governed his thinking: *An Attempt to Explain the Former Causes of the Earth's Surface by Reference to Causes Now in Operation.* "Now" did not mean the present moment, but historical time well-documented by observers across almost two millennia.

No, the pockmarks in the marble were not present during the construction of the temple. By comparison with known marine phenomena, Lyell recognized the holes as the signature of the "date clam" (for its shape), *Lithodomus* (the name means "rock-home"). The date clam burrows into limestone and coral by dissolving the substrate with a mucous-textured, chemical secretion. By means of "clam saliva," *Lithodomus* had etched a bivalve dormitory. These mussels cannot exist unless submersed in seawater; therefore, at some time post-construction, the sea had drowned the pillars.

Lyell ruled out changes in the level of the Mediterranean as the cause. Clearly, rising twenty-four feet, remaining at that level while a host of shelled bivalves etched their way into the columns, then dropping by the same amount, would not have gone unnoticed. Lyell concluded that the culprit was initial subsidence and subsequent uplift. He wrote with confidence that "Antiquity affords, in itself alone, unequivocal evidence that the relative level of land and sea has changed twice at Pozzuoli since the Christian era, and each movement, both of elevation and subsidence, has exceeded twenty feet."[1]

Historical records confirmed not only Lyell's insight, but also a date of uplift in the sixteenth century. Letters penned by firsthand observers told of an earthquake in 1538 that left a terrace of fish out of water that the inhabitants of Pozzuoli easily collected. Lyell searched successfully for analogues to the Pozzuoli events and found several. In 1692 at Port Royal, Jamaica, buildings dropped fifty feet under the sea. In 1822 in Valparaiso, Chile, an earthquake raised the coastal zone several feet. Similarly, the harbor of Penco, Italy, suddenly lifted twenty-four feet in the earthquake of 1750.[2]

Causes Now in Operation

"Causes now in operation" (the key phrase from the subtitle of Lyell's book) in past centuries encompassed volcanism and earthquakes. Lyell elaborated on his interpretation of the pillars at Pozzuoli to persuade his readership of the mobility and mutability of the land. He had identified "causes now in operation" and claimed that such causes could account for earth's structures throughout geologic history. There was no need to invoke "former causes" in remote times (a Noachian deluge, for example) responsible for the Alps, the Paris basin, or any other features of the earth's surface. The subtitle of *Principles of Geology* was a grand statement of extrapolation, now known as the principle of "uniformitarianism," or the aphorism, "the present is the key to the past."

The persuasiveness of his argument hinged in part on the interpretation of an ancient, seaside ruin composed of metamorphosed limestone (marble) communicated to his colleagues via illustration. Other drawings accompanied the text: a sketch of the "ground plan in the environs of the Bay of Baiae" (the top-down perspective) and a cross section (the lateral view) of the marine deposits and cliffs at Pozzuoli. The history of the pillars at the Temple of Serapis unlocked the secret to interpreting the earth. The frontispiece engraving to *Principles of Geology* illustrated what Lyell had conjured, an amalgam of mind and nature.

THE GEOLOGIST'S APPRENTICE

Geology's sorcerer needed an apprentice and, thanks to the gift of his book to a recent Cambridge graduate, he would find the perfect candidate, a grandson of an Age of Enlightenment botanist and poet, Erasmus Darwin. His sisters called the boy "Bobby" after his middle name, Robert. History knows this ambitious scion, anxious to explore the splendor of tropical worlds, as Charles. Courtesy of his favorite Cambridge professor, the Reverend John Stevens Henslow, Charles Darwin received the invitation of a lifetime. Reluctant at first, his father (another Robert) came round to encouraging his son to sign on to the voyage of the *Beagle*.

Old Red Sandstone

Darwin had rather minimal preparation as a geologist, an unlikely profile for someone destined to champion Lyell's ideas. In August of 1831 he trekked through the Welsh countryside with his Trinity College teacher, Adam

Sedgwick, proposer of the Cambrian and Devonian divisions of geologic time. Carboniferous Period rock, bearing strata of coal critical to the industrial revolution, lay above the Devonian. Assuming the age of Old Red Sandstone as Devonian and mapping its distribution helped English geologists prospect for beds of coal.

Hiking on his own, Darwin examined the Vale of Clwyd in a fruitless search for Old Red Sandstone *in situ.* "Fruitless" turned out to be significant—Darwin's field mapping added to Sedgwick's growing body of knowledge about the distribution of layers of Devonian rock in the Welsh mountains, confirmed a broad trough structure (syncline), and led to a substantial revision of the geologic map of the region.[3] Limestone (younger than Old Red) he found aplenty, caverns included, one where a local resident had discovered the tooth of a rhinoceros.[4]

The single-geologic-field-study-seasoned Darwin to join the HMS *Beagle's* second voyage as the ship's naturalist and a gentleman companion to the captain, Robert FitzRoy. The captain had a good reason for wanting a modicum of geologic expertise on the voyage. While mapmaking the ship's first voyage through the islands of Tierra del Feugo, he had observed the influence—presumably, of iron ore deposits—on the bearing readings of the ship's compass. Having a trained geologist with him, he imagined, might prove illuminating—and help locate mineral resources to exploit at some future date.

As Darwin boarded in1831, Captain FitzRoy presented him Charles Lyell's newly released, *Principles of Geology*. Darwin avidly consumed Lyell's work and received shipment of the second volume while in port at Montevideo, Uruguay, a year later. Guided by Lyell's *Principles*, Darwin drew sketches of topography and cross-sectional diagrams as their interpretations, Lyell's notions of subsidence and uplift foremost in his mind.

Darwin undertook his first effort at geologic interpretation upon the *Beagle's* arrival in the volcanic Cape Verde Islands. He enthusiastically sketched outcrops of lava and shelly limestone on Quail Island, then observed that emplacement of these formations was analogous to their occurrence on the larger island, St. Jago. Lava, he decided, had flowed over the seafloor. The island's entire layered structure of lava flows and marine organism detritus subsequently rose. The system worked the same as shown by the pillars in the engraving of the Temple of Serapsis that prefaced Lyell's treatise.

Atoll Fringing Reefs

Carbonate organisms played an essential role in Darwin's geologic thought, and his analysis of coral reefs fringing volcanic islands has stood the test of time. More importantly, his argument about the formation of coral atolls branched from the trunk of experimental science to become a novel, intellectual adaptation to the challenge of inferring the course of events in the distant past.

An understanding of coral atoll formation has acquired urgency in an era of rising sea level, ocean acidification, and increasing temperature. Darwin, the geologist, remains timely.

By the end of the voyage, sailing among the islands of the Pacific, Darwin confidently concluded that "disturbances . . . in the subterranean regions" caused the subsidence responsible for converting a volcano into a lagoon. He speculated that the reefs started growing in the shallow waters on the edge of volcanic islands which subsequently slowly sank into the sea. Meanwhile, the reef continued growing outward. Darwin categorized the coral reefs around Pacific islands as "fringing, barrier, and atoll."[5]

Darwin's *Structure and Evolution of Coral Reefs* "was the first brick laid on the foundations of a revolutionary view of life. It is not just a book about reefs, it is a book which sweeps across the ecology and geology of the whole world."[6] and hypothesized that these three types of coral formations observed in the present are the historical consequence of slowly sinking islands. Sampled at one moment are different stages in a continuous process. One island is an example of another's past, a different one, of its future—places substituting for time to fuel a non-experimental engine of thought, a "second style" of science.[7]

Subsequent studies of atoll formation have supported Darwin's interpretation in some contexts as well as recognized that formation processes have complexities and variations he never imagined.[8] For example, Darwin, similar to Lyell, assumed constant sea level over immense spans of time. His assumption was incorrect—he failed to account for Ice Age glaciation and melting (unknown to him at the time). Over tens of millions of years, coral atolls, moreover, may form on platforms of accumulating limestone (karst) independent of volcanic activity.[9]

The voyage of the *Beagle* revealed to Darwin abundant evidence of subsidence and uplift—past and present—transforming landscapes through time just as imagined by Lyell. For Darwin, the present served as a powerful analogue to the past on both sides of the Andes. At Port Desire, Argentina, he noted a "great *level* plain" extending in every direction "divided by vallies."[10] Oyster and mussel beds on this dry, barren plain meant these strata had once been below sea level. The shells now perched two hundred and forty-seven feet above it—a strong analogy to the story of the second-century Roman pillars at Pozzuoli.

Lifted Mussel Beds

At Talcahuano and Concepción, Chile, he experienced "the most awful yet interesting spectacle" he had ever beheld: a megathrust, subduction zone earthquake.[11] The earthquake helped him to visualize the uplift of the oyster and mussel beds above the Atlantic shore on the opposite site of the continent:

> The most remarkable effect of this earthquake was the permanent elevation of the land. . . . There can be no doubt that the land round the Bay of Concepcion was upraised two or three feet; . . . one little rocky shoal, now exposed, was formerly covered with water. . . . Captain Fitz Roy found beds of putrid mussel-shells still adhering to the rocks, ten feet above high-water mark: the inhabitants had formerly dived at lower-water spring-tides for these shells.[12]

From March through June 1835, Darwin rode across the Andes to Mendoza, Argentina, and back again. On ridges thousands of feet above sea level, he observed beds of ancient seashells. Darwin also found silicified trees, similar to living coastal species, entombed in marine sandstone, now resting seven hundred miles inland. Perception of anomalies in the present guided his visualization of the past.

THE BRANCHING CORALLINE OF LIFE

Darwin's drawings of biological detail and sketches of geologic sections immeasurably aided his thinking. Simply put, Darwin invented imagery for improving the understanding of life. Most any student of biology recognizes the solitary figure embedded in *On the Origin of Species*: the tree diagram that pictures the effects of natural selection.[13] More than two decades before publication of the *Origin*, however, Darwin sketched a figure as prescient

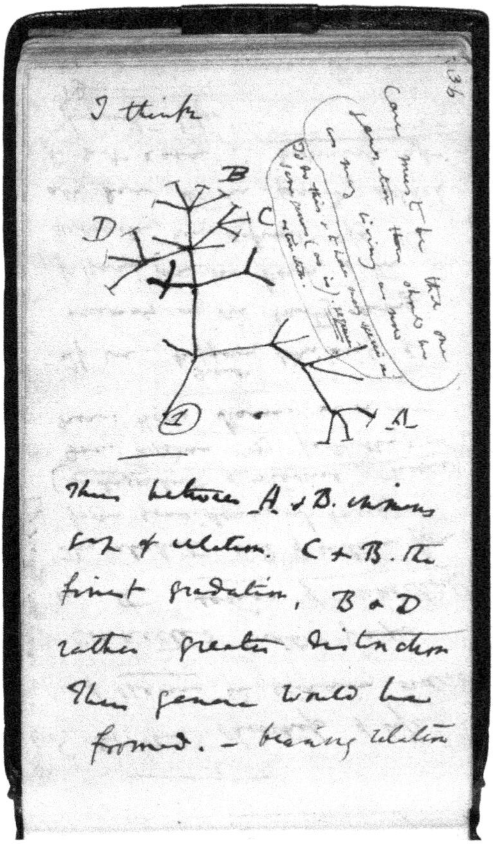

Figure 16.1 Darwin's "I think" Sketch and Entry from His 1837 *Notebook B*. *Source*: MS DAR 121, p. 36. Reproduced by kind permission of the Syndics of Cambridge University Library.

and striking as Einstein's handwritten E = MC^2. In his 1836 *Notebook B, The Transmutation of Species*, there appeared a small, branching diagram with notes that began with "I think . . ." He was thinking about the end points, labeled A, B, and C, on a simple sketch (figure 16.1).[14] His annotations to the drawing read:

> Case must be that one generation then should be as many living as now. To do this & to have many species in same genus (as is) requires extinction.
>
> In his notebook he added:
>
> Thus between A & B immense gap of relation. C & B the finest gradation, B & D rather greater distinction. Thus genera would be formed. - bearing relation to ancient types with several extinct forms.[15]

Red Algae

Even the "I think" diagram had a precursor image: a "coralline" based on the growth pattern of *Amphiroa orbignyana*, a crusty form of sea life collected by Darwin now classified as a red alga (figure 16.2).[16] Overlaying the specimen onto the famous "I think" diagram aligns their branchings to a remarkable degree, as underscored by the German art historian Horst Bredekamp in a book appropriately titled *Darwins Korallen: Frühe Evolutionsmodelle und die Tradition der Naturgeschichte* (*Darwin's Coralline: Early Evolution Models and the Tradition of Natural History*).[17]

In reviewing Bredekamp's book, Florian Maderspacher, senior editor for the journal, *Cell Biology*, found the alignment "charming," but not convincing. Many things branch and spread apart—in shrubby, coralline, or tree-like fashion. Drawing upon historian Julian Voss's critique of *Darwins Korallen*, Maderspacher, pointed out, for example, that drawings contemporary with Darwin's theorizing visualized embryonic development with branching diagrams similar to the "I think" sketch.[18]

Darwin, nonetheless, duly recorded his coralline inspired thought on the transmutation of species: "The tree of life should perhaps be called the coral

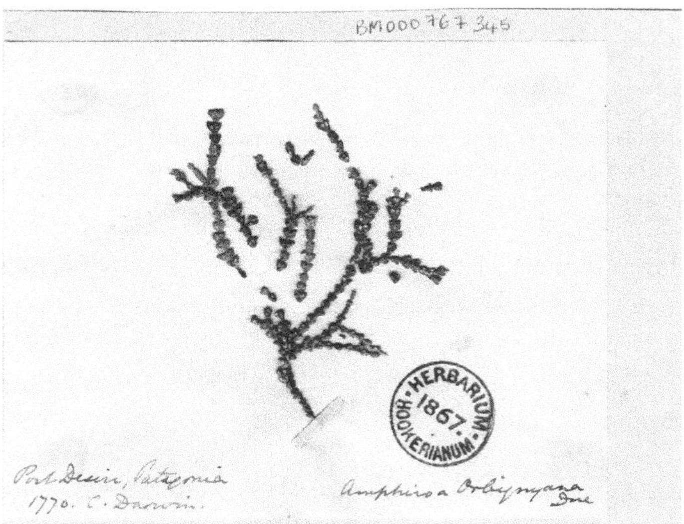

Figure 16.2 The Branching Coralline, *Amphiroa Orbygnyana*, a Red Algae, Collected by Darwin While at Port Desire, December, 1833. *Source*: © Trustees of the Natural History Museum, London. Used with permission.

of life."[19] New calcareous growth—whether polyps or not—grew from the tips of branches. Budding symbolized new life from old; base branches suggested extinct species. The image, despite the confusion of nomenclature (and plant and animal affinities), worked. Branching points indicated common ancestors. Darwin explained that the most recent branchings placed closely related species near to each other—buddings separated by "the finest gradation." The tips of branches far apart stood for an "immense gap of relation."[20]

The branching points stood for hypothesized, common ancestors. Lineages branched and budded new populations like polyps on a coral, variation distancing them from each other and common ancestry uniting them. This insight dominates the pages of *On the Origin of Species*. In its final chapter Darwin returned to the problematical existence of a primordial zoophyte, ancestor to all plants and animals, a question rooted in his mind ever since discovering a leech embryo in a structure misidentified by others as an alga. His "I think" expression had become "I believe":

> I believe that animals have descended from at most only four or five progenitors, and plants from an equal or lesser number. Analogy would lead me one step further, namely, to the belief that all animals and plants have descended from some one prototype. . . . Probably all the organic beings which have ever lived on this earth have descended from some one primordial form, into which life was first breathed.[21]

Reticulated Giraffes

Darwin's branching imagery grew from comparisons of living beings past and present without any knowledge of the mechanisms of inheritance. The tree of life, understood in the twenty-first century, has many tangled branches, announced writer David Quammen after interviewing cutting-edge researchers in molecular biology and genetics. Genes move about in astonishing ways—sometimes from organism to organism, courtesy of retroviruses capable of inserting themselves into a host's genome. Quammen quotes evolutionary biologist Jan Sapp: "A great percentage of our own DNA is of viral origin." As much as eight percent.

How significant are genes moved about by retroviruses? Extremely significant, in both serendipitous or deleterious ways, if inserted into egg or sperm cells. One astonishing example: insertion of viral *syncytin* genes in events scattered across tens of millions of years. Some *syncytins* cause cells to fuse; others suppress immune system responses. As explained to Quammen by Sapp, the function of the placenta depends upon a "fused-cell" layer that supports nutrient and waste transport between the fetus and its mother. *Syncytins* enable cell fusion. Other *syncytins* prevent triggering rejection of the embryo as foreign tissue by the mother's immune system.[22]

Genes, from time to time, scurry from one organism to another in disrespect of species boundaries. Fusion happens, and humans, with insights gleaned from the lifestyles of retroviruses, can engineer genomes. The machinery of deleting and inserting genetic information has "gone viral."

The Darwinian iconography of branching, whether tree or coral, retains much value but no longer suffices, especially among the most ancient origins of living forms. Shrubs and thickets work as transition imagery to improve evolutionary thinking. Intensified branching, however, falls short of the needed break with tradition. The tree metaphor works almost not at all in puzzling out relationships among microbes at the roots of life. The pathways followed by genes through time and bodies crisscross at important junctures. Twigs can reunite—"anastomose" to form reticulating patterns.

The veins of leaves reticulate. The patterning of giraffe hide reticulates. The hyphae of fungal partners anastomose. The strangler fig encircles its host with a confusing, anastomosed network of trunks and branches. Which metaphor works best? "The leaf pattern of life, the giraffe skin of life, the fungal threads of life"? Those icons leap too far. Perhaps "the strangler fig of life," descended with modification from its arboreal ancestor in the struggle for creative insight, works best.

DARWIN'S DEBT

Darwin knew just how indebted he was to Lyell's theorizing about the mutability and transformation of landscapes, in large measure the trigger for appreciating eons of life's analogous mutability and "transmutation" (the term preceding common usage of "evolution"). Change, in both contexts, was due to "causes now in operation," neither supernatural nor restricted to the past. In a letter to Leonard Horner, a Scottish educational leader (and geologist) whose sister happened to be married to Lyell, he admitted:

> I always feel as if my books came half out of Lyell's brains & that I never acknowledge this sufficiently, nor do I know how I can, without saying so in so many words—for I have always thought that the great merit of the *Principles*, was that it altered the whole tone of one's mind & therefore that when seeing a thing never seen by Lyell, one yet saw it partially through his eyes.[23]

The sorcerer and his apprentice bonded well. In their minds' eyes dwelled two marine organisms, one burrowing into limestone pillars, the other secreting carbonate branches. A masterpiece engraving of The Temple of Serapis and a humble sketch of The Tree of Life inspired magnificent thoughts about the mutability of land and life. The images they bequeathed anchored the principle of uniformitarianism and the theory of natural selection and changed humanity's perspective of geologic time.

NOTES

1. Charles Lyell, *Principles of Geology: Being an Attempt to Explain the Former Changes of the Earth's Surface, by Reference to Causes Now in Operation*, vol. 1 (London: John Murray, 1830), *The Complete Work of Charles Darwin Online*, ed. John van Wyhe, accessed June 27, 2020, http://darwin-online.org.uk/.
2. Ibid., 455, 456.
3. Richard D. Keynes, *Fossils, Finches, and Fuegians: Darwin's Adventures and Discoveries on the* Beagle (New York: Oxford University Press), 18–21.
4. Adrian Desmond and James Moore, *Darwin: The Life of a Tormented Evolutionist* (New York: W. W. Norton & Co., 1991), 96.
5. Charles R. Darwin, *The Structure and Distribution of Coral Reefs. Being the First Part of the Geology of the Voyage of the Beagle, under the Command of Capt. Fitzroy, R.N. During the Years 1832 To 1836* (London: Smith Elder & Co., 1842), 48. *The Complete Work of Charles Darwin Online* (2002–), ed. John van Wyhe, accessed June 27, 2020, http://darwin-online.org.uk/.
6. Gordon Chancellor, "Introduction to Coral Reefs," *The Complete Work of Charles Darwin Online* (2002–), ed. John van Wyhe, accessed June 27, 2020, http://darwin-online.org.uk/.
7. Stephen Jay Gould, "Evolution and the Triumph of Homology, or Why History Matters," *American Scientist* 74 (1986): 60–69.
8. Drilling off of the island of Funafuti near the end of the nineteenth century, part of the Royal Society Coral Reef Expedition of 1896-1898, confirmed Darwin's

hypothesis that coral overlay volcanic rock (T. G. Bonney, ed., "The Atoll of Funafuti: Borings into a Coral Reef and the Results," Report of the Royal Society of London Coral Reef Boring Committee [London: Harrison & Sons, 1904], cited in Paul S. Kench, D. Thompson, Murray R. Ford, Hiroki Ogawa, and Roger F. McLean, "Coral Islands Defy Sea-Level Rise over the Past Century: Records from a Central Pacific Atoll," *Geology* 43, no. 6 [June 2015]: 515–518, accessed July 3, 2020, https://doi.org/10.1130/G36555.1). Study of the geology of Tahiti in the twenty-first century corroborated Darwin's explanation of how subsidence may convert a fringing coral reef into an encircling lagoon. The analysis incorporated sea level changes due to cycles of glaciation as well as island subsidence. Reefs climb the slopes of an island volcano while leaving reef terraces behind at former sea levels (Paul Blanchon, Marian Granados-Corea, Elizabeth Abbey, Juan C. Braga, Colin Braithwaite, David M. Kennedy, Tom Spencer, Jody M. Webster, and Colin D. Woodroffe, "Postglacial Fringing-Reef to Barrier-Reef Conversion on Tahiti links Darwin's Reef Types," *Scientific Reports* 4, no. 4997 (May 21, 2014), doi:10.1038/srep04997, accessed August 8, 2018, https://www.nature.com/articles/srep04997).

9. Lakshmi Supriya, "Rethinking Darwin's Theory of Atoll Formation, *Eos* 101 (October 30, 2020), accessed October 31, 2020, https://doi.org/10.1029/2020EO151044.

10. Keynes, Beagle *Diary*, December 24, 1833, 208.

11. Ibid., March 5, 1835, 296.

12. Charles R. Darwin, *The Voyage of the Beagle: Darwin's Five-Year Circumnavigation* (Santa Barbara, CA: The Narrative Press, 2001), 328, ed. Nora Barlow, 2nd ed. (London: Pilot Press, 1945).

13. Charles R. Darwin, *On the Origin of Species by Means of Natural Selection; or, The Preservation of Favored Races in the Struggle for Life*. reprint of 6th ed. (Garden City, NY: Doubleday Dolphin Books, [1872] 1960), 119. In the original Origin published in 1859 by John Murray of London, the figure was a foldout between pages 116 and 117.

14. Niles Eldredge, "Darwin's Other Books: 'Red' and 'Transmutation' Notebooks, 'Sketch,' 'Essay,' and *Natural Selection*," *PloS (Public Library of Science) Biology* 3, no. 11 (November 15, 2005): e382, accessed June 27, 2020, https://doi.org/10.1371/journal.pbio.0030382.

15. Charles R. Darwin, *Notebook B: Transmutation of Species*, (Cambridge: Cambridge University Library, 1837–1838), 36, transcribed by Kees Rookmaaker, *The Complete Work of Charles Darwin Online* (2002–), ed. John van Wyhe, http://darwin-online.org.uk/. Image CUL-DAR 121 © Cambridge University Library, reproduced with permission, http://darwin-online.org.uk/content/frameset?itemID=CUL-DAR121.-&viewtype=text&pageseq=1.

16. Horst Bredekamp, *Darwin's Korallen: Frühe Evolutionsmodelle und die Tradition der Naturgeschichte* (Berlin: Klaus Wagenbach Verlag, 2005). There are three synonyms for this algae *Amphiroa orbignyana*, *Bossea orbignyana*, and *Bossiella orbignyana*. Brederkamp has used "Bossea;" the specimen collected by Darwin and housed in the Natural History Museum, London, is labeled *Amphiroa orbignyana*. http://www.catalogueoflife.org/annual-checklist/2006/show_species_details.php?record_id=18213.

17. Bredekamp, *Darwins Korallen*.

18. Florence Maderspacher, "The Captivating Coral–the Origins of Early Evolutionary Imagery," *Current Biology* 16 (2006): 478, accessed March 17, 2012, https://doi.org/10.1016/j.cub.2006.06.019.

19. Darwin, *Notebook B*, 36–37.

20. Ibid., 25. Darwin's "coral" is not a coral, but a calcareous red alga, ironically named after a rival naturalist, Alcide d'Orbygny, who journeyed in South America just prior to Darwin. Perhaps "The Algae of Life" should replace "The Tree of Life" as an organizer of evolutionary thought.

21. Darwin, *Origin*, 6th ed., 473.

22. David Quammen, *The Tangled Tree: A Radical New History of Life* (New York: Simon & Shuster, 2018), Kindle.

23. Charles R. Darwin to Leonard Horner, August 29, 1844, "Letter no. 771," *Darwin Correspondence Project*, accessed June 27, 2020, https://www.darwinproject.ac.uk/letter/DCP-LETT-771.xml.

Chapter 17

Clawed Wings

Rough sketches and detailed drawings in both woodcuts and metallic engravings advanced natural history across centuries. In 1824, the Royal Geologic Society of London chose to publish illustrations in its *Transactions* using a new printing technique, lithography, developed by Alois Senefelder at the close of the eighteenth century.

Lithography is not an engraving process. Instead, it utilizes chemical solutions to etch and fix exquisite images on rock. The artist draws with grease pencils, then washes the stone's surface to etch some of the limestone unprotected by the greasy drawing. The non-image surface binds to water molecules. A turpentine rinse removes the grease but leaves the area with a coating that repels water. Oily ink adheres to the image area and water to the remainder of the surface. Lithography exploits the repulsion between water and oil at a molecular scale, thus defining precise boundaries. The durability of rock enables the mass production of superb illustrations. Artists continue to work with lithography as an art form (newer techniques use substrates other than limestone rock).

The Royal Geologic Society's adoption of lithography dramatically reduced costs of printing illustrations and ushered in an age of mass production of images of geological and paleontological specimens, landscape drawings, cross-sectional diagrams, and maps. Refinements kept woodcut technology in use as well, another means of cost reduction. For example,

Lyell's 1833 third volume of *Principles* contained many more illustrations than his first two. Lowered expense encouraged lavish productions, crucial to communicating the visual information inherent to geology.[1]

SOLNHOFEN LIMESTONE

For centuries, artisans have quarried fine-grained, Late Jurassic limestone from the Solnhofen region of Bavaria in southern Germany.[2] In the year following publication of Darwin's *Origin*, their toil unveiled a spectacular fossil. Preserved in Solnhofen's limestone, so essential to the lithographer's art, lay the Jurassic enigma, *Archaeopteryx*, and its clawed wing. Every known specimen of this feathered "first bird" has come from Solnhofen limestone quarries. The image of *Archaeopteryx*'s skeleton, its neck arched over its back in the grips of death, rapidly etched itself into Darwinian thought. The fossil was an exemplar of a basic form of life in transition (reptile to bird)—the perfect "missing link" to many.

One hundred years later, Solnhofen's *Archaeopteryx* reprised its center stage role in the drama of reptiles-into-birds drama when Peabody Museum paleontologist and the world's leading *Archaeopteryx* expert, John Ostrom, published more than eighty stippled drawings of bones from the Jurassic dinosaur, *Deinonychus antirrohopus*. The same as Lyell's *Principles*, an influential frontispiece opened Ostrom's monograph. The artist's illustration depicted a creature that would serve a quarter century later as the model for the velociraptors of Jurassic Park (figure 17.1).[3]

Ostrom's raptorian beast, *Deinonychus*, not only inspired popular culture (first the book by Michael Crichton in 1990, then the movies), but also a revolution in the conceptualization of avian evolution. Ostrom inspected the forelimbs of *Deinonychus* and discovered the same unusual "semi-lunate" wrist bone that joined hands to forelimbs in *Archaeopteryx*, suggesting kinship between birds and dinosaurs. In both creatures the hand ended in three, wickedly clawed fingers.

Without the obvious impressions of feathers in the fine-grained Solnhofen limestone that entombed *Archaeopteryx*, fossil skeletons of *Deinonychus* and *Archaeopteryx* would be difficult to distinguish.[4] Dinosaur anatomists now conclude that both species share kinship with tyrannosaurs

Figure 17.1 Robert T. Bakker's 1969 Drawing of the Posture and Gait of *Deinonychus Antirrohopus*, Frontispiece from John H. Ostrom, *Osteology of Deinonychus Antirrhopus, an Unusual Theropod from the Lower Cretaceous of Montana.* *Source*: © Peabody Museum of Natural History, Yale University. Used with permission.

Figure 17.2 *Deinonychus Antirrhopus* Drawn by Armand Morgan, Frontispiece, from John H. Ostrom, *Osteology of* Deinonychus Antirrhopus, *an Unusual Theropod from the Lower Cretaceous of Montana.* Fiftieth Anniversary Edition, 2019. *Source*: © Peabody Museum of Natural History, Yale University. Used with Permission.

and *Coelophysis*, their ancient, Triassic, New Mexico Ghost Ranch cousin whose bronze statue graces the entrance to the Museum of the Earth in Ithaca, New York.

Yale University and the Peabody Museum issued a commemorative issue of John Ostrom's seminal study of *Deinonychus* and its birdy anatomy in 2019. For this fiftieth anniversary edition, a new artist illustrated the frontispiece with a *Deinonychus* sporting feathers (figure 17.2).[5]

FEASTING ON LOCUSTS

In 2002, workers transported a printing press to the Solnhofen quarry so that artist Mario Derra could create a monumental lithograph printed at the site. His work, "Mass Tourism at the Twelve Apostles," measured four-by-forty feet and pictured 100 giant locusts marching in front of a rock formation, the Twelve Apostles. The locusts symbolized a plague of tourists descending upon Solnhofen. The immense locust-etched printing plates reside in the Deutsche Museum together with an exhibit of Alois Senefelder's invention, the lithograph press.[6]

Locusts (*Pycnophlebia robusta*) up to six inches in length can be found fossilized in Solnhofen limestone. They date from the Late Jurassic, the same as *Archaeopteryx*. Perhaps the first bird's feathered appendages functioned as a sweep net to capture giant locust prey and clutch its meal with raptor-like claws.

AVIAN DINOSAURS

Fossils from Spain, China, Mongolia, and Madagascar indicate Mesozoic worlds populated by small, feathered, predatory dinosaurs, the ancestors of birds, scampering after prey, their anatomies pre-adapted for flight.[7] They bequeathed to the earth the exquisite beauty flashing over fields and throughout forests, the extraordinary diversity of form from spoonbill beaks to webbed feet.[8]

The evolutionary insight sparked by Ostrom's drawings of fossil hand bones is now so well established that most paleontologists consider birds to be "avian dinosaurs." "There is a dinosaur outside my window," wrote Steve Brusatte, a tyrannosaur specialist, as he gazed at a seagull.[9]

Science educators at the University of Copenhagen have introduced an activity for secondary school students visiting The Natural History Museum of Denmark that would please Steve Brusatte. The activity accompanies the exhibit, "Evolution: From Dinosaur to Bird" in the Feathers Gallery and challenges students to compare a modern pigeon skeleton to a cast of a fossil *Archaeopteryx*. Based on their observations of similarities and differences, student investigators form hypotheses about the origins of birds and the traits needed for flight.[10]

Many aspects of the evolutionary relationship between dinosaurs and birds—especially the origin of feathers deep in dinosaurian ancestry, for example—remain under debate. Fossils being discovered in China have enriched the discussion. The question, "Was *Archaeopteryx* the first bird to fly?" remains vital and an entry point for students into the practices of paleontology. With modest guidance, they may inspect feather symmetry, fused vertebrae in the tail ("pygostyle"), presence of a sternum, beaks versus jaws with teeth, wing area, and several other anatomical traits for contrasting a modern bird with an ancient, winged creature in fossil form.

The exercise raises a classic conceptual problem, "What good are feathers on a creature unable to fly?" Maybe they helped it to stay warm, or sweep insects into its jaws, or attract a mate with a display of color. The right answer is not the point; a credible line of argument is.

ILLUSTRATING GRASSHOPPERS

"Grasshoppers have five eyes," wrote second-grader Indira in her science journal (years before her study of the Horse Fossil Kit). She next added "and four wings." Beneath her entry she entered the evidence. Indira had carefully pried apart a grasshopper's wings and disassembled its corpse. Using copious amounts of Elmer's Glue, she mounted the legs, wings, and head in her journal. As eyes, she counted with the help of a magnifying glass the three ocelli on the head plus the two obvious, large, lateral compound eyes. Indira knew indeed how to look closely at natural history specimens (figure 17.3).

Indira's grasshopper entry in her science journal combined an authentic natural object with a book, in effect, a micro-museum. One hundred years earlier, science lectures for teachers included a similar exercise.

NATURE STUDY'S GRASSHOPPER

Louis Agassiz, founder of Harvard's Museum of Comparative Zoology in 1859 (the same year as the publication of the first edition of Darwin's *Origin*), influenced legions of public-school teachers in the latter part of the nineteenth century. He admonished them to have children "study nature, not books."[11] Agassiz certainly put his motto into action. To begin a lecture, he handed

Figure 17.3 A Primary Grade Child Glues Grasshopper Parts into Her Science Journal.
Source: Photograph by the author.

out grasshoppers for each teacher to hold and observe. Thus enthralled, they listened intensely to him explaining its body plan and behaviors.[12] He must often have pointed out, "Grasshoppers have five eyes."

In addition to encouraging the teaching of science to children, Agassiz influenced the turn of the century's burgeoning nature study movement predicated upon the priority of visual and sensory experiences. Cornell University's Anna Botsford Comstock, author of *Handbook of Nature Study*, championed Nature Study in the reform of rural education. Comstock emphasized a poetic and aesthetic approach to encounters with nature's familiar objects and creatures, often examined with practical uses in mind.[13] One can easily imagine Indira as her student.

Comstock begins her lesson, "The Grasshopper," by acknowledging that traditional laboratory studies of its anatomical structure place a "burden" on students. Children first pay attention to the creature's jumping, "the peculiarity which makes this insect a 'grasshopper.' . . . Being such a famous jumper, [the grasshopper] does not long burden any object."[14] Outdoors a century later and for thirty seconds, Indira marks each landing with a pebble. She measures the distance of each jump, adds the distances, and determines grasshopper hopping speed.

Her classmates place their still-living grasshoppers in a tumbler to watch them crawl up the sides. They look closely using a magnifying glass and soon discover an oval pad between and in front of the claws. The pad has microscopic hairs that stick to smooth surfaces.[15] In Indira's case, the stickiness is augmented by glue to affix the side of the leg of her once-living grasshopper to the page in her journal.

On the first day of grasshopper study, the teacher dutifully reminded the class that respect for life meant returning captive creatures unharmed to the places where they lived. "Remember, a grasshopper belongs to itself, not you," he said. Alas, he did not realize that the abundance of schoolyard hoppers signaled a year of locust plague. Everyone dutifully released their catch, then, at the end of the day, the crisp crackle of crushed arthropod skeletons could be heard as the school buses rolled up. After that, journaling with grasshopper parts was deemed ethical.

THE GRASSHOPPER'S TOILET

"The Grasshopper" lessons in Comstock's *Handbook* include several black and white illustrations. One fascinating diagram disarticulates seven mouth parts. Another depicts a grasshopper depositing eggs, "the earth partially removed to illustrate an egg-mass already in place" (something to watch for during recess). Of course, there is attention to how a grasshopper sings by rubbing its hind legs together. An unexpected description of another behavior follows the singing lesson:

> The grasshopper makes its toilet thus: It cleans first the hind feet by rubbing them together and also by reaching back and scrubbing them with the middle feet. . . . It cleans the middle feet by nibbling and licking them, bending the head far beneath the body in order to do it. It polishes its eyes and face with the front feet.[16]

Comstock delights in continuing with a description of the grasshopper's "most comical manner of cleaning" the antennae. It firmly plants one down beneath a front foot, then slides the antenna out slowly, then repeats for the antenna on the other side of its head. What a pleasant way to spend a hot afternoon: watching a grasshopper wipe off its antennae with its feet.

FEATURES IN THE CREATURE

Harvard's Agassiz was devoted to excellent teaching, beloved and respected among scholars, and a fierce opponent of Darwinism. He countered Darwin's assertion of common humanity among all peoples with a theory of polygenism: the creation of humans as separate species adapted to geographic climates. Although opposed to the institution of slavery, a number of his writings (especially letters) expressed repugnant racial views.[17] His "centers of creation" imagery, exploited by white supremacists, was a disservice to humanity.

But he knew how to draw and how to interpret landscapes.

According to physicist Edwin Hall, Harvard's Louis Agassiz, the first geologist to recognize the global Ice Age, instructed new students of natural history "to gaze in solitude at a single fish for two or three days." From his prestigious perch atop physics' mathematical paradigm, Hall may not have meant to mock the great Agassiz's dependence on detailed observation, but he likely earned a chuckle from his experimentally rigorous brethren. At least he acknowledged the importance of visual study of life's diversity in the fields of botany and zoology.[18]

There's a true story that prompted Hall's bit of mockery. Samuel Scudder recalled his first meeting in the 1850s with Agassiz, who told him to "take this fish and look at it; we call it a *Haemulon* [scale-finned grunt fish]; by and by I will ask what you have seen."[19]

Scudder looks at the *Haemulon* for ten minutes and believes the task is done. With nothing else to do, he decides to draw the fish. By drawing, he begins to find "new features in the creature,"[20] and soon after proudly shares them with Agassiz, who remains unimpressed. "Look again, look again!" Agassiz says, leaving Scudder "to his misery." A day later he earns a bit of praise. Progress is fine, but Agassiz demands more. "So for three long days he placed that fish before my eyes, forbidding me to look at anything else." Samuel Scudder looked and he looked and he looked . . . until on the end of the third day he turned into a fish himself.[21]

Not really, of course. What Agassiz intended for his students to learn was a method of close looking. Agassiz "implored his pupils to make connections and develop general principles."[22] By looking closely in his own work, Agassiz found correspondence between the succession of fishes through geologic time and their appearances during stages of embryonic growth. Ironically, devotees of Darwinism in Germany seized upon this correspondence as direct evidence for evolution, which Agassiz rejected.

During the first decade of the twenty-first century, English Professor Neal Lerner assisted a biology laboratory class at the Massachusetts

Institute of Technology. As a humanist in the engineer's shop, Lerner's purpose was to break the norm. The topic of study was embryological development of "mutated zebrafish." At first, students reacted in a manner similar to Samuel Scudder who inhabited nearby Boston neighborhoods one hundred and fifty years earlier. They felt reluctant when asked to draw. Unlike Scudder, they believed technology would serve them better and preferred to take digital photos for a PowerPoint. Maybe they could be done in minutes.

Lerner understood Agassiz' pedagogical first principle: draw to learn.[23] Lerner handed out pencils and the drawing proceeded. The Massachusetts Institute of Technology students, to the satisfaction of their instructor, learned to see the anomalous features of the mutated creatures.[24] Like Agassiz, who modeled his message by putting a grasshopper in every teacher's hand, the very model of a modern MIT instructor rejected treating students as passive recipients, vessels for facts, tiredly reciting memorized subjects.

NOTES

1. Martin J. S. Rudwick, "The Emergence of a Visual Language for Geological Science 1760-1840," *History of Science* 14, no. 25 (December 1976), 156–158.

2. Silvio Keller, "Fossils from Solnhofen," *FossNet FossilienStore*, accessed June 27, 2020, http://www.fossnet-fossils.com/solnhofen.html. Romans used Solfhofen limestone as building stone; in medieval times it served well as roof tiles.

3. John H. Ostrom, *Osteology of Deinonychus antirrhopus, An Unusual Theropod from the Lower Cretaceous of Montana*, fiftieth anniversary edition, foreword by Jacques A. Gautheir (New Haven, CT: Peabody Museum of Natural History, Yale University Press, 2019; originally published July 1969).

4. John H. Ostrom, "Archaeopteryx and the Origin of Birds," *Biological Journal of the Linnean Society* 8 (1976): 91–182.

5. Armand Morgan, frontispiece, John H. Ostrom, *Osteology of Deinonychus antirrhopus, An Unusual Theropod from the LowerCretaceous of Montana.* Fiftieth Anniversary Edition. Foreword by Jacques A. Gauthier (New Haven, CT: Peabody Museum of Natural History, Yale University, 2019).

6. Gesa Todt, "Record Lithography: Quarry of Solnhofen Used as Printing Plate," *Litos* 64 (January 2003), accessed June 27, 2020, https://www.litosonline.com/en/article/record-lithography-quarry-solnhofen-used-printing-plate.

7. Kevin Padian and Luis M. Chiappe, "The Origin of Birds and Their Flight," *Scientific American* 278, no. 2 (February 1998): 38–47.

8. Kate Wong, "How Birds Branched Out," *Scientific American* 323, no. 5 (November 2020): 44–49.

9. Steve Brusatte, *The Rise and Fall of the Dinosaurs: A New History of a Lost World* (New York: William Morrow, an imprint of HarperCollins Publishers, 2018), 269.

10. Marianne Achiam, Leonora Simony, and Bent Erik Kramer Lindow, "Objects Prompt Authentic Scientific Activities Among Learners in a Museum Programme," *International Journal of Science Education*, 38, no. 6 (2016): 1012–1035, accessed October 21, 2020, https://doi.org/10.1080/09500693.2016.1178869.

11. Sally Gregory Kohlstedt, "Nature, Not Books: Scientists and the Origins of the Nature-Study Movement in the 1890s," *Isis* 96 (2005): 324–352.

12. Kimberley Tolley, "Study Nature, Not Books: The Nature Study Curriculum 1891–1932," paper presented at the Annual Meeting of the American Educational Research Association (AERA, New Orleans, LA, April 5-8, 1994), accessed June 27, 2020, https://files.eric.ed.gov/fulltext/ED374009.pdf.

13. Audrey B. Champagne and Leopold E. Klopfer, "Pioneers of Elementary School Science: Anna Botsford Comstock," *Science Education* 63 (1979): 299–322.

14. Anna Botsford Comstock, *Handbook of Nature Study* (Ithaca, NY: Comstock Publishing Associates, an imprint of the Cornell University Press, 1911; revised, 1939; fifth printing of the revised edition with a new foreword by Verne N. Rockcastle, 1996).

15. Comstock, "The Grasshopper," *Handbook of Nature Study*, 338–343.

16. Ibid., 340–341.

17. Stephen Jay Gould, *The Mismeasure of Man*, rev. ed. (New York: W. W. Norton & Co., 1996), Kindle.

18. Edwin H. Hall, *Elementary Lessons in Physics, Mechanics (Including Hydrostatics) and Light* (New York: Henry Holt, 1894): iv-v, cited and quoted in John L. Rudolph, *How We Teach Science, What's Changed, and Why it Matters* (Cambridge, MA, 2019), 47.

19. Samuel Scudder, "In the Laboratory with Agassiz," *Every Saturday* 6 (April 4, 1874): 369–370, quoted in Neal Lerner, "Drawing to Learn Science: Legacies of Agassiz," *Journal of Technical Writing and Communication* 37, no. 4 (2007): 381.

20. Scudder, "In the Laboratory with Agassiz," 370.

21. Ibid.

22. Lerner, "Drawing to Learn Science," 380.

23. Ibid., 381.

24. Ibid., 379.

Chapter 18

Pesky Pill Bugs

"Ms. Campbell, I want to make a giant insect," said Reyni.

"A ladybug!"

A short while later Reyni was cutting pieces of construction paper color-matched to the wings, body, and legs of a ladybug beetle. Reyni had scaled a photograph from the nature magazine, *Ranger Rick*, to an enormous size. More than two-feet long, the giant insect replicated the proportions of an actual ladybug. Suspended from the ceiling by monofilament, it attracted everyone's attention. By the end of the week many more giant insects were on display: an ant, a black widow spider, a dragonfly, and, of course, a butterfly. The colorful creations magnified patterns and body parts.

SCHOOL GALLERIES

The secret to accurate enlargement was the use of a grid. Children placed a transparent grid of small squares over an original illustration or photograph of an insect. On a much larger sheet of paper they drew a grid of squares several times bigger than the ones on the transparent sheet. In each large square they drew by hand (rather than using a pantograph) the part of the insect observed through the corresponding square on the small grid. The process guaranteed proper leg attachment to the thorax, not the abdomen. Cutouts from finished drawings were traced on construction paper and assembled with glue and tape. For the next few weeks, the fourth graders invited their friends to stop by and visit the "giant insect museum" their classroom had become.

When Ms. Campbell moved to central New York she seized upon children's interest to create another museum exhibit at school, this time in an unused storage closet for a gallery. It was spring and the smelt were running in Cayuga Lake. Lake life became the focus on the exhibit, with smelt the star attraction. Several members of the class created a background mural in tempera paint. With guidance from the art teacher, others fashioned fish, crawdads, and eels to suspend in front of the mural. A team of fourth-grade writers authored guidebooks (index cards with a ring to hold them together) to the exhibit. Once completed, children from the primary grades toured the Fish Museum. They enjoyed listening to the interpreters—in several cases, their older siblings.

Projects such as the Giant Insect and Fish Museum integrate art and science. Making art part of a science lesson enhances motivation because the arts provide multiple ways for students to express their understandings while keeping them attached to emotions.[1]

THE MUSEUM OF THE EARTH

High above Cayuga's waters in central New York state stands the venerable Paleontological Research Institute, a repository since the 1930s of Devonian trilobite fossils and much, much more. In 2003 its sister institution, the Museum of the Earth—dedicated to the fusion of art and paleontology for public audiences—opened. Cecil, the statue of an early theropod dinosaur (*Coelophysis*), greets each entrant to the Museum of the Earth.

Cecil the Coelophysid

In late Triassic time (on the order of 200 million years ago), near present-day Ghost Ranch, New Mexico, flash flooding buried a pack of these pony-sized precursors to velociraptors. New Mexico chose *Coelophysis* as the state dinosaur. Georgia O'Keeffe, famed artist of the landscapes of New Mexico, made Ghost Ranch her home.[2] The symbolism of union between art and science represented by *Coelophysis*' Ghost Ranch pedigree equals the beast's evolutionary significance. Coelophysis represents the lineage leading to *Tyrannosaurus rex.*

Coelophysis left three-toed tracks on the other side of the continent in New York's Triassic muds. New York had to make do with footprints, its lonely claim as home to a dinosaur, while New Mexico boasted of abundant skeletal fossils. The footprints, nevertheless, earned *Coelophysis* the status of a museum mascot in bronze.

Past Cecil a stunning mural, painted by Barbara Page, fronts the ramp to the main exhibit hall at the Museum of the Earth.[3] Five hundred and forty-four panels, one for each million years of earth history from Cambrian to Pleistocene time, depict fossil plants and animals in situ and brilliant color, all at the same scale. The Cambrian trilobite, *Paradoxides gracilis*, graces the mural's 524th panel. At fifteen inches, *Paradoxides* was a big trilobite, among the largest ever found. Trilobites, sporting wraparound, compound eyes and strong exoskeletons with flexible joints, flourished for hundreds of millions of years.[4] Eurypterids (sea scorpions) preyed upon them.

Pipsqueak Armadillos

The flexible-joint body plan of the Arthropoda has endured and flourished—among insects, spiders, ticks, lobsters, and, of course, pill bugs. Children enamored of pill bugs' armadillo-like segmented structure will recognize its truly ancient body plan clad in arthropod armor, the same as trilobites.[5] Pill bugs ("roly polies"), along with lobsters and true crabs, are bonified crustaceans (trilobites are in a class by themselves). Pill bugs are small, terrestrial isopods ("iso" meaning "equal" and "pod" meaning "foot"—each leg ends with a foot the same as on every other leg). Mucking around in gardens turns up both pill bugs and sow bugs from a sister family, the Onisidae. That sow bug family name is forgettable, but the pill bug one is a hoot: Armadillidiidae.

Trilobites are collectable, extinct, fossil arthropods. Pill bugs (and sow bugs) are living, collectible arthropods, popular in science projects (they fit in the nose cones of model rockets and easily survive parachuting back to the ground) but disliked by gardeners. Able to grip and crawl along kite string, they perform high-wire acts at school as science circus performers. Some marine species of isopods parasitize fish by latching onto their gills or tongues. This form of parasitism concerned William Shatner, the actor who played James Kirk in the *Star Trek* series, in an episode of *Boston Legal* about investing in fish-farming. Shatner's character, Denny Crane, humorously referred to the isopods as "Klingons."[6]

X-rayed Muscles

There's no lack of trilobites in the Museum of the Earth galleries. One in particular exhibits a remarkable degree of preservation. Its skeletal structure, muscle attachment, and digestive track all appear on display—plus some guesses as to what food was still in the gut at the time of its death, approximately 450 million years ago (late Ordovician Period). These extraordinary details are known courtesy of Cornell University Professor Emeritus John Cisne's X-ray photographs of *Triarthrus eatoni*. Because his *Triarthrus* kept quite still, Cisne was able to use long exposure times that resolved details, such as muscle attachments and stomach contents.[7]

The two X-ray photographs of trilobites hanging side by side at the Museum of the Earth are duplicates of the ones above Professor Cisne's living room sofa. They are art and science at one and the same instant. In fact, visit any scientist's home or office or lab. There is certain to be a piece of art reflective of the scientist's love of subject.

IMAGES MAY BE HARMFUL TO THE IMAGINATION

Representations of evolution in scientific art merit cautionary labeling. For example, in Rudolph Zallinger's magnificent mural, *The Age of Reptiles*, at Yale University's Peabody Museum, *Tyrannosaurus rex* stands featherless and erect, like an actor in a rubber Godzilla suit. Jurassic Park's *T. rex* aligns her spine less vertically to accommodate a better understood bipedal gait, but still reflects Zallinger's archetype. In 2020, a child dressed in a *T. rex* costume for Halloween necessarily stands in an upright posture, but thank goodness, at least knows to deck out the costume with a bit of plumage.

March of the Hominids

Zallinger reinforced the imagery of evolution as foreordained and ladder-like in illustrations for Time-Life's 1965 book *Early Man*. Zallinger drew fifteen figures in a four-page foldout portraying a step-by-step progress from knuckle-walking *Pliopithecus* to confident, upright-striding *Homo sapiens*. Zallinger's art echoed the frontispiece to Thomas H. Huxley's *Evidence as to Man's Place in Nature* published in 1863 in the wake of Darwin's *On the Origin of Species*. Huxley's illustration compared living great apes—gibbon, orangutan, chimpanzee, gorilla, human—implying progress (toward shorter arms, for example).

Zallinger's image (and its Huxley antecedent) became part of "America's visual vocabulary," parodied and satirized seemingly without end. Matt Groenig, creator of *The Simpsons*, drew a panel of the evolution of "Homersapiens" with Homer carrying a banana.[8] Another iconic image matches Zallinger's march of the hominids (and prehominids): Ersnt Haeckel's drawings comparing the embryological development of fish, salamander, turtle, chick, pig, cow, rabbit, and human embryos at three different stages of development.

The Embryo's Tale

By greatly reducing the separation of humans from beasts, Haeckel's drawings shocked his contemporaries. In their early stages, Haeckel demonstrated, humans were beasts—or at least virtually indistinguishable from chicken embryos. He drew imprecisely, perhaps biased by his intent to promote the idea that the development of an organism proceeds through stages corresponding to their descent. Some accuse him of outright fraud and others of carelessness.[9] He incorrectly insisted that embryonic stages resembled ancestral adults rather than just recognizing similarity of development as evidence of common ancestry.

The analysis of developmental homologies (same structure, same relative position in different embryos) has clearly contributed to evolutionary biology. Haeckel's images of embryos, however, continue to worm their way into present-day classrooms. The Teacher Institute for Evolutionary Science has assembled an extraordinary collection of resources, among them a link to a classroom activity from Public Broadcasting System's Nova series: "Timing is Everything." The activity presents scrambled drawings of fish, pig, chick, calf, and human embryos in three stages of development. The task is to order them by species and then compare similarities and differences, realizing how similar all five are in the earliest stages.[10] It does not, however, link these forms to ancestral adults, as did Haeckel.

Five months before Darwin's death and shortly after publication of his treatise on worms, *Punch's Almanack* published a cartoon satirizing him.

The whiskered evolutionist clad in a toga, watches as a parade of primates unfolds. From one creature to the next tails shed and postures "improve." The first creatures, limbless and embryonic in appearance, slither. The caption reads, "Man is but a worm."[11]

GUIDING IMAGES

Imagery often guides thinking. Imagery can stimulate novel ideas and improve insights. To quote John Dewey, "Science states meanings; art expresses them."[12] Yet artful imagery does more than express meaning. Art amplifies intelligence by expanding the possibilities of thought and perception. Applying Eisner's "criteria of artistry" (chapter 14) helps to understand science and improve curriculum design. Science, like art, makes "the obscure vivid" and "aesthetic experience and credibility possible." Scientists and artists focus attention by constructively neglecting certain features of phenomena in order to discover truths about others.[13]

Aesthetic expression punctuates advances in the history of paleontology in particular and geology in general. Framed by these disciplines, there is beauty in an evolutionary view of life, its several guiding images descended from a provocative few—an engraving of a Roman marketplace, a sketch of a coralline alga, a limestone fossil of *Archaeopteryx*, a detailed drawing of a fish, a close look at a grasshopper, a high-crowned tooth.

Evolution's images in both scientific and popular media have entertained and teased for an exceedingly long time. The educational trick for teaching science is to promote imagery that contributes to improved understanding while being watchful lest the uncritical acceptance of an image confuse or misguide. Guiding imagery, whether anchoring metaphor or actual drawing, fosters an aesthetic sense for, and appreciation of, a subject, two essential and pleasurable dimensions of understanding.

NOTES

1. Emily Clare Riggins, "Integrating the Arts and Sciences in the Museum Setting," Master's thesis (Seton Hall University, 2014), accessed August 29, 2020, https://scholarship.shu.edu/dissertations/2039.

2. Steve Brusatte, *The Rise and Fall of the Dinosaurs: A New History of a Lost World* (New York, NY: William Morrow, an imprint of HarperCollins Publishers, 2018), 65–67.

3. *Rock of Ages, Sands of Time*, paintings by Barbara Page, text by Warren Allmon (Ithaca, NY: Paleontological Research Institution in association with The University of Chicago Press, 2001).

4. Ibid., 16.

5. Robin Burnett, *The Pillbug Project: A Guide to Investigation* (Washington, DC: National Science Teachers Association, 1992).

6. *Boston Legal*, season 2, episode 3, "Finding Nimmo," produced and written by David E. Kelley, directed by Mike Listo, aired on October 11, 2005 on ABC. William Shatner speaks two very humorous lines. When an environmental lawyer seeks his support for protecting wild salmon, Shatner's character, Denny Crane, responds, "I came out here to enjoy nature. Don't talk to me about the environment." Back at the hotel, his law partner, Alan Shore, played by James Spader, informs Denny that lice (parasitic isopods), nicknamed "cling-ons," spread by farmed fish can kill wild salmon. Denny looks up from reading and says, "Did you say Klingons?"

7. John L. Cisne, *Triarthus Eatoni (Trilobita): Anatomy of Its Exoskeletal, Skeletomuscular and Digestive Systems* (London: Forgotten Books Classic Reprint, 2018).

8. Geoffrey Giller and Richard Conniff, "Iconic. Almost by accident," *Yale Alumni Magazine* 78, no. 2 (Nov/Dec 2014), accessed July 16, 2020, https://yalealumnimagazine.com/articles/3977-march-of-progress.

9. Nick Hopwood, *Haeckel's Embryos: Images, Evolution, and Fraud* (Chicago, IL: University of Chicago Press, 2015). Haeckel's drawings of embryos from his 1874 *Anthropologie* appear on Hopwood's cover. Debate over the fraudulent nature of these drawings stems from embryologist Michael Richardson's (St. George's Hospital Medical School, London, England) suspicions. Richardson found discrepancies between rates of embryonic development across species and Haeckel's drawings synchronized by stage by comparing them with photographs. Richardson argued that Haeckel exaggerated similarities and paid no attention to scale (the drawings are all similarly sized). See Elizabeth Pennisi, "Haeckel's Embryos: Fraud Rediscovered," *Science* 277, no. 5331 (September 1997): 1435a, doi:10.1126/science.277.5331.

10. "Odyssey of Life, Part I—The Ultimate Journey," classroom activity accompanying *Timing is Everything*, November 24, 1996, NOVA Online, accessed August 17, 2020, https://www.pbs.org/wgbh/nova/teachers/activities/2317_odyultim.html.

11. Linley Sambourne, "Man Is but a Worm" *Punch's Almanack for 1882* (*Punch*, December 6, 1881). Accessed March 27, 2013, Wikimedia Commons, http://commons.wikimedia.org/wiki/File:Man_is_But_a_Worm.jpg.

12. John Dewey, *Art as Experience* (New York: Berkley Publishing, Penguin Books, 1934; Perigee edition, 2005), Kindle.

13. Elliot W. Eisner, *The Kind of Schools We Need* (Portsmouth, NH: Heinemann, 1998), 152–153.

Chapter 19

Flashy Plumage

While crossing the Atlantic in January of 1832 on board the HMS *Beagle*, Charles Darwin collected plankton with a net towed behind the ship. The precocious young naturalist (just twenty-two at the time) wrote, "Many of these creatures so low in the scale of nature are most exquisite in their forms & rich colors. It creates a feeling of wonder that so much beauty should be apparently created for such little purpose."[1] Maybe the purpose is too obvious. The beauty exists to share.

To teach science is to share beauty and create wonder. Scientific inquiry values skepticism, of course; scientists are professional skeptics. Skepticism helps prevent them from being hornswoggled. In equal, or perhaps greater measure, scientists value the experience of wonder. Nature's beauty, closely observed, never fails to astonish, to cultivate wonder.

AESTHETIC EVOLUTION

"How has the 'aesthetic complexity of nature' come into being?" asks ornithologist Richard O. Prum in *The Evolution of Beauty*.[2] Prum's primary topic is the male-female coevolution of bird plumage, song, and courting rituals—primary examples of what Darwin referred to as the results of "sexual selection." But he also interprets humanity's devotion to the fine arts as a product of coevolution. Composer and audience, artist and critic—producers and evaluators of aesthetic entities—coevolve. Mozart, for example, composed symphonies "that transformed his audiences' capacity to imagine what music could be and do."[3] Cultural preferences for creative works that give pleasure, inspire thought, or transform meaning shape the evolution of aesthetic expression.

Prum generalizes sexual selection as "aesthetic evolution." Of course, Prum's view is contentious, but there is no escaping the compelling nature of his topic of interest: beauty. Beauty is in the mind of the beholder, whether of a trilobite appendage, bird plumage, or a symphony. Beauty rules.

Coevolution by Choice

For Prum "coevolution" is neither metaphor nor analogy, but, rather, a process shared by humans and nonhumans whenever choice exists. By making sexual, social, and ecological choices, "animals design themselves."[4] The peacock's sexual ornamentation coevolves with the peahen's aesthetic preferences. Changes in plumage and preference for changes reciprocate. The same correspondence exists between preferences and behaviors such as birdsong.

Preferences accelerate the cycle. As change gathers momentum, remarkable aesthetic productions evolve. Prum stresses the examples of the complex dance of the male manakin and the elaborate structures built by male bowerbirds, in both cases driven by female, sensory, and cognitive judgments. The effect of myriad individual mating choices thus designs displays and ornamentation. Aesthetic evolution in this sense has piggybacked on natural selection ever since the origin of choice, conscious or otherwise. Evolution, ever opportunistic, waits for chances to repurpose traits and fashion unanticipated novelty.

Late Jurassic fossils of *Anchiornis huxleyi*, a remarkable four-winged creature that lived over 150 million years ago, and several other species from the Liaoning Province of northeast China have convinced paleontologists that the dinosaur ancestors of birds had feathers. Even the "melanosomes" (cell organelles that absorb light) that colored their feathers have fossilized in abundance. Comparing them to the melanosomes of existing birds has enabled the reconstruction of most of *Anchiornis*' plumage in full, colorful, and iridescent glory.[5] Apparently, flashy plumage preceded flight—perhaps put to use by dancing dinosaurs in courtship displays.

Alas, these reconstructions probably render the computer graphic animations of featherless velociraptors and *Tyrannosaurus rex* anachronisms. They likely resembled *Anchiornis* in many feathered respects. Perhaps a return to the mammal quarries of the John Day Formation may yield fossilized melanosomes revealing whether Oligocene horses were striped like zebras, spotted like Appaloosas, or neon pink (with sparkles) like a child's stuffed unicorn. Even if beauty may fail to rule, it always fascinates.

The Ubiquity of Aesthetics

Aesthetic appreciation is basic to the sciences. Particle physicists respect the beauty and elegance of their theories; cosmologists nearly lose their souls to the Hubble telescope's deep space images. Ornithologists dwell on the elegance of feathers. The inescapable aesthetic pleasures of geoscience range from micrographs of thinly sectioned mineral crystals illuminated with polarized light to multicolored maps and cross sections of continental collisions. No compelling interest in science exists apart from an aesthetic appreciation of a particular phenomenon.

Disciplines embed artistry in their practices. Especially in the geosciences, representing data, theorizing, and communicating depend upon visualization and spatial reasoning. The task is to see through the landscape and creatures of the present into the features of the past that created them.

Aesthetic pleasure dwells in the study of life's history and in the reconstruction of ancient landscapes. The same sense of pleasure belongs in the science classroom. Its expression deserves cultivation as an embodiment of learning equal in importance to cognitive outcomes. Aesthetic principles guide scientific practice in the present and artistic expression has contributed to its advance. Good science teaching helps students improve upon their imagery for making sense of the natural world and aesthetic criteria guide understanding to a feeling of satisfaction.

THE POWER OF ICONOGRAPHY

"Iconography is a neglected key to changing opinions," notes science historian and paleontologist Stephen Jay Gould in *Wonderful Life*. Whether for ill or noble intent, he continues, "Every demagogue, every humorist, every advertising executive, has known and exploited the evocative power of a well-chosen picture."[6] Mileposts in the history of scientific imagery, especially of ancient life, may remind educators of this principle so that they might use it wisely.

Scientific art has the potential to instill appealing misconceptions able to endure for generations. The evolution of the horse did not climb a "ladder" of progress; a hard hoof is not an improvement over splayed toes. The hoof expresses the nature of grassland habitat; toes, the forest floor. The branching image of a "tree" does injustice to the complexity and diversity of horse family creatures through time. Stephen Jay Gould emphasizes the need to view the evolution among horses as "bushy."[7] Rather than a triumph of perfection, the modern horse is a lonely survivor from a once diverse lineage. The diversity of the antelope family across Africa and Asia, by contrast, still exists. Today there are many ways of being a successful antelope. The lineage of the horse as a multibranched shrub flourished in the past.

Oversimplification of the story of horse evolution invites incredulity, if not derision. Creationists find strong evidence for "gaps" in the fossil record in the wildly improbable story of five successive species evolving from *Eohippus* to *Equus*—the image of a single trunk. The jumps are just too big. Embedded in textbook teaching, the toe-to-hoof, browser-to-grazer scenario has invited evolution's skeptics to mock it.[8]

Anti-evolutionist publications juxtapose the foot of the "dawn horse" (*Eohippus*) with that of a modern thoroughbred to prove that transition is impossible. The simple image of progress, unless dispelled, increases susceptibility to Darwin denialism. The John Day Fossil Beds National Monument Horse Fossil Study Kit (quite a name), the star of chapter 11, countered the stereotyped, linear-ladder progression with lessons that emphasized variation within fossil collections through time and place, methods matched to the task of documenting populations in transition.

TOOTH HISTORY

Being a dentist is a risky business for a mouse, especially when a fox with "a rotten bicuspid and unusually bad breath" appears at your office.[9] William Steig's tale of Dr. De Soto, the mouse dentist that usually avoids treating predators, has entertained children for several decades. Wry, understated storytelling carries the narrative, but Steig's illustrations pack an even greater punch. In combination, each amplifies the other: picture a fox with its mouth propped open, expelling awful fumes, as a crane lifts Dr. De Soto into position to extract the rotten bicuspid. The story is delicious, the art sublime.

The Horse Fossil Study Kit draws attention to teeth—to premolars and molars rather than bicuspids—and the stories they have to tell about evolution and ecology. From start to finish, the kit relies upon visualization. Understanding the evolutionary significance of skull and teeth fossils depends upon perception of forms and patterns. In combination, geologic story and fossil imagery amplify each other. Absent visualization, the study of earth history feels barren.

The curriculum immerses adolescents in the practices of paleontology by featuring stunning replicas of life's exuberant past. Every lesson about discoveries from the John Day Fossil Beds relies upon visual imagery: tooth crowns, eye positions, skull shapes, data charts. At critical junctures, complex diagrams depicting adaptations and migrations represent evolutionary descent. Reconstructing the regional landscape, transformed by volcanic activity through geologic time, challenges imagination.

Professional artisans and artists joined with amateur sculptors (the students who fashioned clay-action figures) to create the instructional resources. The program valued authentic phenomena; its sequence cohered around their images. The curriculum, furthermore, attached itself to the land the students inhabited and to the field research sites their teacher had visited. Truths can be found where tooth fossils lie.

Achilles' Heel

Despite the omnipresence of visual materials and visual exercises, the curriculum rarely called upon students to draw—or animate, or sculpt, or photograph. They observed, measured, categorized, sequenced, and plotted data derived from fossil horses—the norm for a secondary school science lab. The exercises helped them to construct a conception of variable populations within a charismatic mammal family adapting to a dynamic landscape on a geologic timescale. Instruction resisted the call upon students to mirror the aesthetic dimensions of the Horse Fossil Kit's resources with their own aesthetic expressions of understanding.

Of course, time is the ruling tyrant of school science. Seldom does a class have time to complete all of the lessons in the kit, not to mention the extensions in the appendix. States and districts mandate what biology teachers must follow. Aesthetic appreciation and artistic expression do not hold a prominent place in science standards. In many disciplines, they do.

Everything Paleontological

The Horse Fossil Kit used in Ms. B.'s biology class integrated topics of ecology, climate change, and evolution. Tangible artifacts and simulations using real-world data focused upon the importance of finding viable intermediate forms in the transition of a lineage from ancestral to modern genera. Exercises engaged students in scientific reasoning and argumentation in the style of paleontologists.

Ms. B. used the fossils, in addition, to underscore the concept of genetic variation (the unit that preceded the study of evolution), subject to selection pressures. She felt that the class appreciated the opportunity to experience the flavor of paleontological research about "a cool animal, one they were familiar with."

The shifting climate of John Day Country through geologic time mirrored the present-day divide between a temperate rain forest zone to the west and a region of high desert to the east of the Cascades (the volcanic mountain chain in the Pacific Northwest). The kit organized investigations of particular interest to students who lived in the region. Students reconstructed scenes through time from their own backyards. The science was timely, particular, and local—an example of teaching science beyond standards.

Seldom does paleontology hold prominence in school science. Nevertheless, Warren Allmon, director of the Paleontological Research Institution in Ithaca, New York, reminds everyone that paleontology holds keys to understanding the earth's climate as well as the story of life's evolution. The records of life are the records of ocean circulation and geochemical cycles, of seismic events and extraterrestrial object impacts. Allmon exaggerates, of course, when he claims that "everything is paleontology." At the same time, the field consistently connects large-scale thinking to tangible events through artistic imagery.[10]

NOTES

1. Charles Darwin's *Beagle Diary*, ed. Richard D. Keynes (Cambridge: Cambridge University Press, 2001), January 11, 1832, 22.

2. Richard O. Prum, *The Evolution of Beauty* (New York: Anchor Books, a division of Penguin Random House, 2017).

3. Ibid., 344–345.

4. Ibid, 345.

5. Ibid., 152–153.

6. Stephen Jay Gould, *Wonderful Life: The Burgess Shale and the Nature of History* (New York: W. W. Norton & Co., 1989), 28. "Iconography" is the branch of art history that interprets the content of images.

7. Stephen Jay Gould, *The Structure of Evolutionary Theory* (Cambridge, MA: Belknap Press of Harvard University Press, 2002), 580–581.

8. Paul W. Kroll, "Evolution Gets the Horse Laugh," *The Plain Truth* 34, no. 11 (November 1969): 23–29.

9. William Steig, *Doctor de Soto* (New York: Farrar, Straus, and Giroux, 1982), 12.

10. Warren D. Allmon, "Everything Is Paleontology," *Paleontological Research Institution Blog* (January 16, 2020), accessed October 5, 2020, https://www.priweb.org/blog-post/everything-is-paleontology?rq=warren.

Part IV

CONCEPTUAL COHERENCE

Chapter 20

Poetic Rocks

A series of "blue sky days" greeted dignitaries arriving in Beijing during the November 2014, meeting of the Asia-Pacific Economic Cooperation forum. The region's economic leadership breathed the unusually clean air, courtesy of President Xi Jinping's curtailment of Beijing's smokestacks and tailpipes. Temporary rules closed factories, restricted traffic—license plates ending with an odd number driving on one day, plates with an even number on the next.

TEACH LESS/LEARN MORE

That same day, several graduate students were crossing the campus of Beijing Normal University not far from Tiananmen Square to participate in an international gathering of a different sort. Enamored with the power of networks in systems science and linguistics, Beijing Normal University quantum physicist and systems scientist Wu Jinshan had searched for a research paradigm in education to counter what he felt was the all-too-frequent leading role of rote learning in China.

Professor Wu valued grasping and clarifying the "big picture" over factual recall and formula memorization. Settling upon the approach to meaningful learning pursued by Joseph Novak and his protégés,[1] Jinshan named his system "Teach Less/Learn More" and assembled an international team to demonstrate teaching for conceptual and creative thinking in the sciences.[2]

The instructors aimed to teach less while expecting students to learn more by emphasizing a subject's typical ways of thinking, characteristic methods of analysis, and connections to other fields.

A Teach Less/Learn More plenary session in meaningful learning preceded subject-themed workshops such as Thinking Geologic Thoughts.[3] The participants, representing a host of disciplines (e.g., geography, fine arts, mathematics, neuroscience, education, linguistics, physics, nuclear science) enrolled according to personal interest. Could they, over the course of two or three weeks of evening workshops designed according to Novak's ideas about improving education, acquire meaningful insights into an unfamiliar subject? Would they find the subject conceptually coherent or impenetrably abstruse?

Where to begin? With art, of course—because absent the training of perception, no one feels inclined to think geologic thoughts.

FROM KOYAANISQATSI TO GONGSHI

The art and architecture pervading the campus beckoned attention. En route to the workshops students passed numerous Gongshi—"scholar's stones" or "spirit rocks," some with inscriptions—fronting buildings, enlivening gardens, lining walkways.[4] Gongshi pleased perception, fostered contemplation, and teased interest. Decorative rock, polished to reveal intricate granitic textures, clad nearby buildings. Grasp of geology's skeleton of relationships, its interpretive and analogical spine, its limbs of temporal and comparative logic, and its hard-to-crack challenge of scale in every bone, promised enriched encounters with campus rock.

Chinese tradition honors rocks as expressions of beauty and pattern, wisdom and reason. Director Godfrey Reggio and producer Francis Ford Coppola's film, *Koyaanisqatsi*, captures these values as expressions of landscapes on the Colorado Plateau. In his film, scenes of monoliths in Monument Valley and canyons unfold in a hypnotic sequence of color and form: Zion, Grand, and Bryce. The title, a Pueblo Indian word, means life in turmoil.[5]

Coppola contrasts the serenity of rock with frenetic commercial and industrial life: traffic, refineries, mines. Ironically, from a geologic perspective, rock also records turmoil. Rocks speak of a cascade of change that never achieves balance, of a dynamism on scales both vast and sudden. Processes of uplift and subsidence, erosion and sedimentation, interfere to truncate or accelerate each other, yielding the landscape of the moment. The accumulation of the past is quite beautiful to behold.

Thinking Geologic Thoughts opened with the sights and sounds of *Koyaanisqatsi*. Later, attention returned to Beijing Normal University's Gongshi. In between fell lessons on interpreting landscapes changing on a geologic scale, leveraged by analogy to ice. Ice's buoyancy, structure, and density presage factors that sculpt the earth's crust.

IS ICE A ROCK?

The workshop's first lessons paired the seriousness of ice formation with the whimsy of a moose's taste for logical argument, a strange start for learning how to interpret the earth. Analysis of ice as rock, especially its formation during winter, spans the chasm separating everyday perspective from realization of how seafloors may become mountain tops or sink toward the core of the earth.

Freezing Thoughts

Decades ago, geologist William Romey posed a disarming question about a familiar substance. He simply asked, "Is ice a rock?" A child might answer, "No ice is ice and rock is rock. Rocks are really hard. And ice is slippery." Ice is hard—but ice breaks. Rocks break, too. Sometimes they are slippery.

What might set the boundary between the meaning of "rock" and the meaning of "ice"? How are they similar to, but different from, each other? Rocks sink and ice floats. Is that always true? Is that a sufficient criterion for telling them apart? The class pondered this question as the instructor annoyingly rejected each reason presented for doubting the "rockness" of ice.

A rock may begin as a liquid, then congeal when cooled. It freezes into a crystalline structure, a frozen concoction of minerals for geologists to hammer on. Ice, too, forms from a chilled liquid, a brittle material for bartenders to split with an ice pick. Rock and ice share a pattern (solids composed of crystals) to process (freezing) at a basic level. Rocks melt into lava. Classifying ice as rock leads to a disquieting inference: water is lava. Stereotypically, rocks consist of several minerals, each with its own freezing point; water is just one "mineral" and hence has just one freezing point. Of course, quartz rock consists of one mineral, silica, plus some minor amounts of others.

Similar implications continue to unsettle settled understanding. Many kinds of sediment sufficiently glued together form rock. Falling snowflakes blanketing the landscape stand as a clear example of a sedimentary process. Cemented together, perhaps by a freezing rain, snow turns into layers of sedimentary rock.

Extreme pressure may alter the mineral composition and fabric of rock, whether formed from a liquid melt or layers of sediment, earning the new composition the appellation of "metamorphic rock." In the case of ice, the metamorphic condition goes by the name "glacier." The analogy continues, centered on more subtle correspondences between ice and rock and exemplifying principles of thinking geologic.[6]

Frazil Ice

Romey exploits winter ice and snow as models of rock formation and landscape features with the complexity expected of students studying to become professional geologists. "Frazil" ice—crystals that form and agglomerate in supercooled, turbulent water—raft together to "roof" a river, a process that resembles crystal formation in a mass of moving lava. Does standing by a river mean having the opportunity to witness a lava flow? Apparently, yes.

Initially, crystals forming in a freezing mud puddle form lengthy fingers. These elongate crystals entrap water that freezes later, but not as fingers (look closely at a shallow mud puddle as it starts to freeze to notice this process). The pattern, explains Romey, characterizes the texture of igneous rocks that form from magma congealing underground into basalt, for example.[7]

Carried by rivers, frazil ice accumulates as icebars—analogues to sandbars around obstructions to water flow. Icebars display the same geometric patterns observed in flows of lava and magma. As these accumulates grow into sheets, they exhibit rippled surfaces and fractured blocks that resemble Hawaii's ropey "pahoehoe" (pa-ho-ee-ho-ee) and blocky "aa" (ah-ah) flows.[8]

Movements create ridges and domes; domes crack and water seeps through, freezing in patterns the same as intruding and extruding lava on a volcano's slope. Inside lava tubes lavacicles grow drip by drip from hot, liquid rock, the same as icicles from drips of icy, cold water. "Hot" and "cold" are human-centric terms of perception. From the point of view of rock and ice, forming icicles is what happens at a temperature close to freezing. There's really no meaningful difference freezing water and congealing lava at the level of molecular organization. Both are phase transitions.

Winter Ice Tectonics

The current of water beneath the surface moves and distorts the sheets of ice; slabs spread apart, drift, and collide. Currents of water differing in density due to temperature (and in the ocean by salinity) well upward, sideways, and sink, jostling the slabs above, analogous to plate movements. Buoyancy keeps ice afloat; it creates a lift force opposite to gravity based on density differences. Ocean floor rock, initially swollen by heat, chills and shrinks in volume, changing its density with respect to the semifluid rock beneath it. It then sinks, somewhat like an ice cube in alcohol, liquid less dense than water.

Fortuitously, for common sense there is an eruptive pattern that separates the flow of ice at freezing temperature from lava. Lava is sticky and sticky means resistant to flow (high viscosity), water flows "like water," not syrup. Erupting underwater, freezing, and cracked by pressure, lava keeps oozing out to congeal in puffy, bulbous forms known as "pillows." When soft, they are much too hot to lay your head down. When cool, much too hard. But pillows they are, in geologic dialect.

Moose Logic

Icy analogical talk, exploratory and playful in nature, risked overloading conceptual circuits. The method in this madness morphed for Beijing Normal students into a task to construct clear, counter, and borderline cases of a familiar category—rock, for instance—then compose criteria for case membership (or nonmembership). When teaching according to a metacognitive strategy, whimsy helps to temper philosophical seriousness, as Morris the Moose, a children's book character, exemplifies.[9]

Morris is, even if a silly moose, a very logical one. Anyone (at least anyone listening to the voice of their inner child) hearing his story cannot escape from wondering, "Is a cow a moose?" Before addressing the question, "What makes a rock a rock?" the class contemplated the conundrum, "What makes a moose a moose and not a horse or a deer or a cow—or is a moose some sort of cow, or deer, or horse, or none of the above? On what basis?"

The whimsical story is quite simple. Morris goes about his daily business, munching on greens, while happening to meet other four-legged creatures. To the first he encounters, Morris comments, "You are a funny-looking moose." The creature replies, "I am a COW. I am not a MOOSE!" Morris takes exception to this claim, explaining, "You have four legs and a tail and things on your head," the same as does Morris. Therefore, he reasons, "You are a moose."[10] The self-proclaimed cow does not appreciate being called a funny-looking moose; with the exception of Jessica, a Vermont dairy cow wooed some years ago by her wannabe mate, Bullwinkle, cows tend to find moose antlers rather unbecoming to a cow.[11] On occasion, a female moose may find a Vermont horse attractive.[12]

Morris reasoned: if a creature is a moose, then it has four legs, a tail, and things on its head. The creature he encountered had all these features. Morris must be forgiven his fallacy. He affirmed the consequent—four legs—and on this basis inferred the antecedent—mooseness. True, moose have these traits. True, the creature he met one day had these traits. But as the reader recognizes, the cow is a cow and not a moose. Many animals have things on their head, four legs, and a tail; not all of them are moose. Cow and moose similarities and differences, in terms of shared traits, are useful elements of playful logic the same as "rock" and "ice." They are also the raw data for drawing evolutionary inferences, as are ice and rock for connecting geologic processes and patterns.

Befuddled by Morris' argument, the cow attempts to negate his conclusion with a new fact: "But I say MOO!" Quickly, Morris moos in response. No crucial difference there. In exasperation the cow shouts, "But my mother is a COW!" As her mother's child, she must also be cow. Cows birth cows. Morris disputes her heresy. "You are a MOOSE! So, your mother must be a moose, too!"

Morris and the "cow" go in search of an independent adjudicator of their dispute. They find a "deer." The cow observes the deer's antlers, an unmistakable mark of mooseness, and thus believes the deer to be just another moose and therefore potentially biased. When they meet a horse, the horse greets them, asking, "Hello horses, what are those funny things on your head?"[13] They stop to reflect upon their images in a still pond of water, where they recognize and quickly accept their differences. The horse is a horse, of course, and a talking one at that.

Case Thinking

Moose logic brings to focus the utility of clear, border, and counter-case thinking to grasping the meaning of a concept. Deer (the borderline case) are remarkably similar to moose (the clear case example). Both belong to the deer family, Cervidae. Cows are a bit less related (another borderline case), and horses are perhaps something else altogether (a counter-case example).

Morris paid no attention to hooves and toe numbers in his construction of family membership (horses come odd-toed, cattle and deer, even-toed). Horses, the true outlier, have no head armor (except in the case of unicorns). Deer sport antlers—bony material, shed annually, well-supplied with blood during growth. Cattle have true horns—life-lasting sheaths of tough protein (the stuff of fingernails, keratin) atop bony protuberances from the skull. Therein lie the criteria for distinguishing category membership to a first approximation: antler, horn, and toe, the directions for their growth inscribed at the genetic level. In the case of antlers, possibly by a cancer related gene.[14]

Human-Centered Meaning

"Come on! Dive in!" shouts Morris as he prepares to leap. "I'd rather not; it's lava," answers the cow. No one wants to dive into lava; it's too hot for comfort and too thick to swim through.

Humans wishing to connect geologic patterns to earth processes invent categories useful for their explanatory purposes. How materials behave at the typical temperatures at the surface of the earth, within its mantle, or on distant bodies of the solar system, calls for meaningful distinctions. Viscosity distinguishes silica melts from water. Another property does, too. Water, as ice, is a perfectly good rock, but, unlike most rocks at earth surface temperatures, it can also be a prominent vapor.

Messing about, even conceptually, cannot be done in haste. Concepts serve purposes. Geologic concepts differentiate—vapor pressure, viscosity, density, buoyancy; sedimentation, crystallization, recrystallization—to boost the mind's capacity to navigate the natural world of ice and rock. Even adult learners need time to juggle their thoughts while struggling with new ideas.

THE PLATE TECTONICS STORY LINE

To rephrase the Walrus from Lewis Carroll's *Through the Looking Glass*, "The time has come to talk of many things: of plates—and quakes—and sinking slabs, of tectonics and time. And why the seafloor tumbles underground, and how the continents decline." "Plates" metaphorically expresses the nature of the earth's crust: a collection of mostly rigid crustal rock jostling about through geologic time. "Tectonics" attempts to explain their structure and formation, the forces that move them, and how the material beneath the crust affects them.

Four Geologic "Stories"

The central characters of density, buoyancy, and pressure appear in most every plate tectonic story. Mostly, plates float. Somewhere, edges of plates sink. In other places magma rises to the surface to birth new crust and stitch plates together in a crazy quilt pattern. Buoyant plumes from the mantle can break them apart. These processes and patterns unfold analogously to the formation of winter ice. To grasp their role is to grasp the "big picture" of plate tectonics theory meaningfully.

There is subtle order in the seemingly haphazard distribution of the remnants of earth's past, in some places on magnificent display. The syllabus introduced the essential organizer of modern geologic thought, the theory of plate tectonics, and wound through four regional interpretations of plates colliding and plateaus lifting:

1. The battering of the Indian subcontinent into the belly of Asia.
2. The destruction of seafloor beneath North America's Cascadia shores.
3. The accumulation and excavation of the rocks of the Grand Canyon of the Colorado.
4. The lifting of the Colorado and Tibetan Plateaus and their associations with changing climates.

Gongshi and Jingshi

In the midst of study of the four geologic stories came, in essence, the "midterm" exercise: "Find a Gongshi on campus that appeals to you. Take its picture. Plan to share with your peers something special about this rock." The task fused aesthetics, emotion, and cognition. Students told how Gongshi evoked feelings, memories, and knowledge. Properties of hardness, smoothness, color, and form gave the stones value as works of art in themselves, sculpted by sedimentary, igneous, and metamorphic processes.

The silence of stone speaks of time, its vastness and its passing. "Each rock is a moment of time . . . at each instant, the summary of its past and the threshold of its future."[15] Even a monumental gravestone is as ephemeral as the span of a lifetime. "A drawing cut in obsidian fades."[16] There are truths, the chants of eternity, to think about while gazing at a rock. Expressions of wisdom and gratitude in stone cross cultures, as expressed in a Chinook prayer of praise etched by artist Maya Lin in the steps of the Art Walk trail at Cape Disappointment State Park in Washington state:

> We call upon the earth our planet home, with its beautiful depths and soaring heights, its vitality and abundance of life, and together we ask that it teach us and show us the way.[17]

The class walked past a statue of Confucius to enter the lobby of the Jingshi Hotel perched on a corner of the campus. Stairs of porphyritic granite graced the entrance. Slabs of travertine, migmatite, and gneiss covered the walls. Polished limestone, replete with fossils, floored the elevators. The sink counters in the restrooms were cut from a dark gabbro. Behind the concierge desk was a wall of brecciated marble. The lobby functioned as a field site for the "final exam."

For each type of rock, students stood close and looked for patterns at the smallest scale the unaided eye could detect. They then stepped back to take in a larger section and again searched for patterns. Finally, they gave each material a creative name that could help to recall its characteristics: melted ice cream for migmatite, shiny ice for marble, crystal soup for porphyritic granite, black sugar candy for gabbro, pulled taffy for gneiss. Apparently, several members of the class had a sweet tooth.

A two-fold interpretive task ensued: imagine a process responsible for fashioning each kind of rock, then speculate how, in the context of plate collisions, such rock might possibly form. Where might the igneous, sedimentary, and metamorphically formed rocks of the Jingshi Hotel lobby be encountered along a north to south transect from Tibet into India? Plausible answers, not necessarily correct ones, received accolades.

The tour ended in the hotel elevator. The class rode together in small clusters, their attention focused on fossils embedded in the floor. They assembled at the top where the instructor paraphrasing, John McPhee, shared the stunning fact that marine limestone, miles above sea level, capped the highest summit on earth.

The Jingshi elevator, when stopped at the hotel's top floor, laid claim to being the highest elevation on campus. There, too, lay a slab of marine limestone.[18]

NOTES

1. Joseph D. Novak, *Learning, Creating, and Using Knowledge: Concept Maps as Facilitative Tools in Schools and Corporations* (Mahwah, NJ: Lawrence Erlbaum Associates, 1998).

2. Wu Jinshan, "Teach Less, Learn More at Being Normal University," 1st Asian Pacific Conference on Concept Mapping (CMC-AP 2017), Yingdong Conference Hall, Beijing Normal University (BNU), Beijing, People's Republic of China, September 20–22, 2017.

3. Joseph D. Novak and Alberto J. Cañas, "The Theory Underlying Concept Maps and How to Construct and Use Them," Technical Report IHMC Cmap Tools 2006-0 (Institute for Human and Machine Cognition, 2008), accessed November 18, 2019, https://cmap.ihmc.us/docs/theory-of-concept-maps.php.

4. The School of Life, "The Wisdom of Rocks: Gongshi," *The Book of Life*, The Wisdom of Rocks: Gongshi, accessed May 11, 2020, https://www.theschooloflife.com/thebookoflife/the-wisdom-of-rocks-gongshi/.

5. *Koyaanisqatsi: Life Out of Balance*, directed by Godfrey Reggio, produced by Francis Ford Coppola, music by Philip Glass (Santa Monica, CA: Metro Goldwyn-Mayer Studies Home Entertainment, 2002).

6. Stephan Gregory Custer, "Snow as a Field-Teaching Medium for Earth Science," *Journal of Geological Education* 39, no.1 (January 1991): 36–37.

7. William D. Romey, "Winter Ice and Snow as Models of Igneous Rock Formation," *Journal of Geological Education* 31 (1983): 9–12.

8. Ibid. The direction quotation is "Once the surface of a river or lake is iced over there are remarkable similarities to the surface features of lava flows."

9. Charles R. Ault, Jr., *Do Elephants Have Knees? And Other Stories of Darwinian Origins* (Ithaca, NY: Comstock Associates, Cornell University Press, 2016).

10. Bernard Wiseman, *Morris the Moose* (New York: HarperCollins [1959] 1989), 5–7.

11. Pat A. Wakefield and Larry Carrara, *A Moose for Jessica* (New York: E. P. Dutton, 1987).

12. Associated Press, "Horse, Moose Become Friends in Vermont," *USA Today*, accessed August 30, 2004, http://www.usatoday.com/news/nation/2004-08-30-moose-cow_x.htm.

13. Wiseman, *Morris the Moose*, 25.

14. Elizabeth Pennisi, "Cancer Genes Help Deer Antlers Grow Bone," *Science* (June 20, 2019), doi:10.1126/science.aay4620.

15. David Leveson, *A Sense of the Earth* (Garden City, NY: Doubleday Anchor Natural History Books, 1971), 129.

16. Xavier University, "Native American Prayers: Only for a Short While (Aztec)" *Jesuit Resource*, accessed May 7, 2020, https://www.xavier.edu/jesuitresource/online-resources/prayer-index/native-american.

17. Maya Lin, *Confluence Library*. The Confluence project memorialized in landscape-embedded art moments of encounter between the Corps of Discovery (the expedition led by Merriweather Lewis and William Clark, 1804-1806), and Native Peoples of America. Lin is the daughter of Chinese immigrants to the United States. She designed the Vietnam Veterans Memorial, a two acre work of stone within the Constitution Gardens of the National Mall in Washington, DC, accessed May 7, 2020, https://www.confluenceproject.org/library-post/cape-disappointment-state-park/.

18. Student names in this chapter have been changed to preserve anonymity. Only first names appear, selected from a list of common names in China. The chapter compresses three workshops (fall semesters at BNU, 2014, 2015, and 2017) into a single story.

Chapter 21

Vast Moments

Early in the 1991 movie, *Grand Canyon*, the character Simon, played by Danny Glover, tells Mack (Kevin Kline) that standing on the rim of the canyon makes "you just realize what a joke we people are. What big heads we got thinking that what we do is gonna matter all that much. Thinking our time here means diddly to those rocks. It's a split second we been here, the whole lot of us. And one of us? That's a piece of time too small to give a name."[1]

From a different perspective, "the piece of time too small to give a name" holds inestimable value, and, in fact, it does have a name: "now," the moment in which consciousness exists and with consciousness the responsibility to make choices—choices that can reverberate on the geologic timescale. The present moment matters, pregnant with possible futures, some of which human action might harm. Author John McPhee's phrase, "deep time," aligns a sense of vast durations with an understanding of the earth right now. Indeed, from the rim of the Grand Canyon time actually looks deep.

Geoscientists love places in the field—the chaotically disrupted orderliness of outcrop patterns, the majestic scenery hiding secrets of landscapes long departed, the record of comings and goings of extinct beasts under changing climates. Places sample earth's history; the present Grand Canyon is a mosaic of traces of the past.

Historically, geologists debated whether the earth remains in a steady state of destruction and renewal or follows an endless, irreversible trajectory.[2] Crustal plates have assembled into supercontinents and rifted apart cyclically on the scale of hundreds of millions of years. Living organisms, the passengers on the plates, have evolved continuously into unique forms. Cycles may abound, yet irreversibility triumphs.

DEEP RESPECT FOR THE PRESENT MOMENT

"Fathoming deep time is arguably geology's single greatest contribution to humanity," claims geologist Marcia Bjornerud, in her book, *Timefulness: How Thinking Like a Geologist Can Help Save the World.* She coined the word "timefulness" to mean "a clear-eyed view" of humanity's role on earth from the perspective of geologic time. Bjornerud, wisely rejects commonplace strategies for communicating human insignificance in the context of geologic time even as she counsels appreciation for its vastness. When representing time since the origin of the earth from fingertip to nose on an outstretched arm, for example, trimming a fingernail wipes out human history. On the scale of a twenty-four-hour day representing four and one-half billion years, recorded history spans a fraction of a second.

> This is a wrongheaded, and even irresponsible way to understand our place in Time. . . . It suggests a degree of insignificance and disempowerment that not only is psychologically alienating but also allows us to ignore the magnitude of our effects on the planet in that quarter second.[3]

The proper way, argues Bjornerud, to think of humanity's portion of earth history is to ask, "What happens after midnight?" Accepting the geologic timescale as a great gift to humanity inevitably raises a nonscientific question. What is the value of understanding time on a geologic scale? The midnight moment holds supreme importance for the future. The image of a time line hinders this thought.

Whether on a trail along the rim of the Grand Canyon or on a yardstick kept in a schoolroom closet, human history, relative to geologic time, lays claim to just .004 percent of the length. Such a tiny percent at first glance seems trivial yet echoes a common practice for teaching about geologic time: representing time as length. A time line is, of course, a preeminent example of how geoscientists represent time. Perhaps it is the most dominant example, at least in schools. Time, to the contrary, is not length—length is just one way of metaphorically representing duration keeping scale intact so that comparisons of relative duration may be made.

Four-thousandths percent of geologic time verges on insignificance. Is the duration of .004 percent of earth's history, on some scale, of equal importance to the other 99.996 percent? Or is .004 percent of geologic time supremely important though minuscule? Should the representation of geologic time as length humble or inspire a student? Choosing and interpreting time's representation faces the inescapable question, "What is the value of understanding deep time?"

In order to answer, imagine walking a geologic time line such as the 1.8 kilometer Trail of Time at the Grand Canyon representing 1.8 billion years, or simply a set of posts marking geologic periods, scaled as lengths, from the Proterozoic to the present in a large schoolyard. Walking forward in time appears conducive to illustrating the cumulative nature of events—the past becoming the present moment.

Representing geologic time as length should counter, not reinforce, feeling insignificant when confronted by the reality of time's vastness. The proper aim of constructing a time line is to cultivate *deep respect for the present moment*. From this perspective, the past accumulates; the present is the sum of past events. As time passes, the importance of the present increases, for what has gone on before continues to exercise influence and shape the future. Time increases like an expanding bubble, its surface an eternal, liminal moment named "now." As William Falkner poignantly wrote in *Requiem for a Nun*, "The past is never dead. It's not even past."

Along the south rim of the Grand Canyon of the Colorado River winds the Trail of Time for 1.8 kilometers. At regular intervals quarried boulders rest in the manner of classic sculpture on plinths. Ordered by age, the rocks sample each formation of rock in the canyon from the ancient Elves Chasm gneiss to the canyon topping Kaibab limestone. Beside each one stands a tube, fixed in angle and direction for viewing the same rock on the wall of the canyon below the north rim. The trail offers aesthetic pleasure and invites field investigation. It is a museum of both science and art, a coherent story and experience.[4]

In the last scene of *Grand Canyon*, the film's key character, Simon, peers down the chasm from the south rim trail, and says about the struggles in life, "I think it's not all bad." The place has a knack for making visitors feel that way.

COHERENT IDEAS

In retreating from the dominating paradigm of universal science, philosopher Stephen Toulmin advocated paying attention to the "nature of the case." He has remonstrated against the quest for criteria of scientific certainty independent of subject, history, and culture and elevated the stature of timely, local, and particular science.

Geologic inquiry, a clear case of Toulmin's position, responds to the challenge of interpreting complex systems on multiple scales in time and space. The objects and events of interest are historical entities. Singular statements reflect their contingent nature. Anomalous associations suggest common cause; convergence of independent lines of investigation on the same cause adds to its credibility. Determining sequence, synchrony, and duration constrains, confirms, or rejects geologic claims. Analogues fuel comparisons and contrasts; metaphors boost insights. Artistic expression contributes to conceptualization and communication.

Examining the contexts and warrants for claims about meteor impacts, caldera eruptions, ghost forests, or seafloor mountaintops provides rungs for climbing to the heights of disciplined geologic thought where concepts, representations, principles, strategies, theories, and criteria of ideal explanation cohere. To make this coherence one's own empowers learners to witness reality through the minds of many others, then act on this understanding with responsibility.

The Worth of an Idea

A good curriculum reveals the usefulness of ideas—their worth—in the context of inquiry disciplined by both thought and method. Sensing the power of ideas is an emotional experience. Understanding feels satisfying once knowledge has demonstrated its worth in achieving an explanatory purpose.[5] Use of ideas in the context of inquiry establishes their meaning; value in achieving distinct aims, their significance. In context, ideas are purposeful, inviting, accessible, authentic, and valuable. Important ideas become a body of knowledge, by interlocking with each other to represent experience, reveal meaning, and demonstrate worth.

Important concepts are deeply layered with meaning, understood best by their use in multiple, explicit contexts. Awareness of a concept's purpose and recognition of its value are essential to feeling the satisfaction of meaningful learning. Learning begins with paying attention and proceeds by grasping meaning, then organizing and reorganizing understanding in response to new meanings. The grasp of meaning encounters significant difficulty when the scale or occurrence of events exceeds the experience of everyday perspective, as in geology. Often, the task of the learner is to grasp "exceedingly unobvious" meaning; the responsibility of teaching is to make unobvious meaning less so.[6]

From an emotional point of view linked to cognition, coherence stems from a sense that one's knowledge is adequate to the problem, that understanding is practical, that the effort to learn will prove productive. Such wide-awakeness to learning attaches emotion to the improvement of understanding. For students in the habit of meaningful learning, cognition and affect cohere to offer strength and recovery in the face of the often troubling and confusing demands of thinking. On the other side of this challenge awaits a feeling of pleasure anchored to conceptual coherence.

Conceptions of geologic phenomena and practices of geologic inquiry respond to the distinctive challenge of time's vastness. This vastness coupled to the short duration of human life presents a fundamental challenge to the conduct of geologic inquiry by scientists and the achievement of geologic understanding among novice students. Deep respect for the present moment—whether grasped through the aesthetic appreciation of the Grand Canyon or felt as concern for the rising risk of unprecedented drought—teaches the value of responsibility for the future, an ultimate aim for learning geologic ideas.

People derive a sense of geologic time from comparisons that underscore the enormity and number of changes in the history of the earth. Achieving such knowledge pleases many persons because the meaning of time gives a sense of one's place in the universe. Rock speaks to the geologist metaphorically, deceiving in its eternity, thrilling in its message of change, and enigmatic in its shadowy preservation of the past in the present.

The Capacity for Renewal

In geologic time the earth displays its capacity for renewal, a message of hope. Too often modern frustration regarding human impact on natural systems devolves into a lament about "lost Edens"—a sense of resignation that the best that can be hoped for amid expanding populations and growing consumption is to slow the rate of degeneration of the natural world. This impoverishment of imagination serves humanity poorly. No landscape is the true Eden; nature recuperates.

Only in the present moment does the opportunity to execute responsibility exist and only for creatures who know time consciously as past, present, and future. Nature's productions have utilized immense spans of time; present landscapes and living creatures encode much of that history. Knowledge of deep time leads to humility and respect for creation—and, ideally, these values lead to good choices about the future. The concept of geologic time in a personal context reinforces what makes humanity human.

Humanity's role during .004 percent (and soon more) of geologic time matters to the protection of species and conservation of resources from the present moment forward, for an immeasurable span of time yet unborn. Celebrating the earth's capacity to rejuvenate, to remain young, to bring forth in abundance, to age with spectacular grace, honors the myth of Eden in a proper way, argues David Oates, author of *Paradise Wild* (2003).[7] Never is Paradise literally lost. Eden remains a place in myth, mysteriously capable of inspiring hope, a hope that children may feel when cultivating a garden, restoring a stream bank, belling the cat to protect backyard birds—or contemplating with wonder the Grand Canyon from its South Rim along the four and one-half kilometer Trail of Time representing four and one-half billion years of earth history.

To appreciate geologic time means to enter the landscape with the question of change foremost in mind, to see through present landforms, structures, and materials and question what might have come before as well as what might come next. The value of understanding geologic time resides in its lesson of human responsibility, a responsibility anchored in humility toward, and respect for, the natural world. Ideas about geologic time cohere best when learned within a context of value. A visitor stands on the rim of the Grand Canyon gazing at a landscape deceitfully frozen in time. The walls stare back, as if to ask, "What comes next?"

NOTES

1. *Grand Canyon*, directed and produced by Lawrence Kasdan, written by Lawrence Kasdan and Meg Kasdan (Los Angeles, CA: 20th Century Fox Studios, 1991).

2. Stephen Jay Gould, *Time's Arrow Time's Cycle* (Cambridge, MA: Harvard University Press, 1987).

3. Marcia Bjornerud, *Timefulness: How Thinking Like a Geologist Can Help Save the World* (Princeton, NJ: Princeton University Press, 2018), 16, Kindle.

4. Ryan Crow, Karl Karlstrom, Laura Crossey, Steve Semken, Deborah Perry, Michael Williams, and Judy Bryan, "It's About Time: Innovations in Geoscience Education at the Grand Canyon, *Legacy* 22, no. 1 (January/February, 2011).

5. Stephen Toulmin, *Human Understanding: The Collective Use and Evolution of Concepts* (Princeton, NJ: Princeton University Press, 1972). In Toulmin's evolutionary metaphor for understanding, "concepts" vary within "populations" and are subject to "intellectual selection. He addresses the problem of conceptual change, finding novelty and continuity. Concepts are individually grasped, grow in meaning through use, and survive by demonstrating their worth in achieving "explanatory ideals." To be rational, in Toulmin's sense, is to commit oneself to the improvement of understanding—to conceptual progress.

6. Charles R. Ault Jr., "The Everyday Perspective and Exceedingly Unobvious Meaning," *Journal of Geological Education* 32, no. 2 (1983): 89–91.

7. David Oates, *Paradise Wild* (Corvallis, OR: Oregon State University Press, 2003).

Chapter 22

Storied Geology

Beneath a drawing of a pipe, the French surrealist artist, René Magritte, wrote, "*Cenci n'est pas une pipe.*" The translation is "This is not a pipe."[1] The caption and image combine to mean that the drawing represents reality; reality is not the drawing.

Disciplines depart from each other in how to represent features of reality deemed salient to their respective agendas. What counts as "real" in one field may not exist in another's representations: metaphorically, different disciplines draw different "pipes." Students, in turn, study such representations to draw reality in their minds. Inescapably, they bring to bear ideas proven useful in everyday life as well as prior knowledge. Sometimes ideas that cohere in the realm of personal experience conflict with a discipline's realities, resulting in disinterest or misconceptions. Conflicting thoughts struggle to cohere and achieving coherence advances understanding. Ideas cohere with reality by means of representation.

GEOLOGIC REPRESENTATION

Alfred North Whitehead's elegant insight, "Understanding is useful," guided Jinshan Wu's "Teach Less/Learn More" program at Beijing Normal University. Whitehead called for education "to impart an intimate sense for the power of ideas, for the beauty of ideas, and for the structure of ideas."[2] The Thinking Geologic Thoughts workshop introduced students to several interpretations of regional geology. Each interpretation invoked a small set of principles to guide thinking and introduced the forms of representation typical of geologic argument. Stories of plateau uplift, plate margin tectonics, and Grand Canyon stratigraphy provided context.

Geologists, the same as other scientists, represent their observations in forms responsive to the characteristic challenges of their discipline. Carefully arranged and analyzed, the facts of observation become artifacts of knowledge that function to amplify patterns in the data. The tasks of learning are (a) to scrutinize these artifacts in search of their structure, (b) to seek out and internalize their coherence, and (c) to determine the purposes they serve. As expressed by John Dewey, a subject "needs to be *psychologized* (Emphasis in orginal); turned over, translated into the immediate and individual experiencing within which it has its origin and significance."[3]

Thinking geologic thoughts parallels thinking paleontological ones: both acquire coherence through visualization reliably communicated. Successful geologic inquiry depends upon rendering large-scale phenomena into observable artifacts: geologic maps, stratigraphic columns, block diagrams, and cross-sectional profiles, for example. The conception of time, a fundamental reality difficult to comprehend, permeates geologic representations.

Since its inception, geologic inquiry has depended upon exquisite illustration to invent and share ideas. Visual representations render temporal relationships on vast scales tractable to thinking. The epitome of geologic illustration, the two-dimensional geologic map, conveys how the topography of today intersects the landscapes of the past. Mapping reveals the transformation of landscapes in three dimensions through time. Colors and lines (bold, dashed, toothed) indicate materials, their ages, and the types of boundaries where they come into contact. Symbols and values, for example, "strike and dip," record the spatial orientation of important surfaces and structures.

Transect lines cross maps to slice open the topography and extrapolate the structures beneath the surface, the buried pages of geologic story. Crustal elements—ocean floors and continents—fragment, migrate, sink, rise, fold, shatter, melt, flow, congeal, and erode . . .

> Glaciation, river scouring, rock compression, and outcropping all contribute to the topography [of a landscape], but do so on very different time scales. . . . Topography [landscape] is the interference pattern between differently scaled processes.[4]

The landscape is, at its essence, the interference pattern left by a host of processes—from tremors to tempests—as they unfold across a range of scales in time and place.

Ubiquitous stratigraphic cross sections tell the sequential geologic history of a location. Correlation diagrams signify synchronous epochs in widely separated places. Series of block diagrams—affectionately referred to as "cartoons"—communicate causal explanations and often compare competing models of regional landscape and geologic structure formation.[5]

Those who summit Mt. Everest stand on snow-covered seafloor. In a series of block diagrams, geologist and mountaineer Mike Searle explained how rocks destined to become the peak of Mt. Everest began as seafloor sediments, became pressure-cooked subterranean fare, and later enjoyed uplift and exhumation.[6] His block diagrams captured stages of collision, faulting, and metamorphosing

THE NATURE OF GEOLOGIC INQUIRY

The present contains remnants of the past distributed as confusing puzzles. Geologic inquiry depends upon concepts and practices responsive to the challenge of interpreting the earth—connecting the patterns of its landscapes and structures to the processes that formed them. Geoscientists strive to solve these puzzles with methods of inquiry adapted to this end. Often unable to execute experimental methods, geologic inquiry stands as emblematic of diversity among scientific ways of knowing.

Geologists value comparing ancient landscapes to modern analogs, substituting place for time, determining temporal relationships, and entertaining multiple working hypotheses. Confidence in geologic solutions increases when independent lines of evidence converge on the same explanation and solutions to problems on different scales cohere. Properties of the phenomena of interest to geologic inquiry are contingent on their history and statements about historical entities make claims about distinct patterns ("singular statements" with limited generality) in contrast to universal laws.[7]

Causal Solutions

The interpretation of a landscape requires solving problems at different scales. At a minimum, the explanation of crystal formation within local rocks must not contradict the account of regional crustal movements. Persuasive arguments in geology demonstrate a common cause accounting for patterns observed on different scales.

Consider how a large meteor impact at the edge of land and sea disturbs the earth. The impact blasts a crater on a regional scale, shatters crystals at a micro one (shocked quartz), launches fragments of melted rock into the wind that chill to become glass beads (tektites) raining down upon the regional scale landscape. The impact pulverizes rock, generates heat, ignites fires, sends a surge of water—a monumentally scaled splash—far inland, and darkens the skies with clouds of ash that circle the globe for months.

As a result, extinction rates skyrocket, not only regionally but globally. The fallout of ejecta from the impact, including an isotope of an element rarely found in concentration on earth but more abundant elsewhere in the solar system (iridium), find their way into sediments at distant locations.[8] Absent a common cause, this set of events remains a puzzling anomaly. They cohere if pattern and process conform to a common cause and dating indicates their sequence and synchrony harmonize with this cause.

Carefully sequenced and synchronized, traces of past events often create a set of anomalous associations across time and place. Doubting an explanation of coincidence, a geologist seeks a plausible common cause rendering these associations into an expected pattern on multiple scales.[9]

Temporal Reasoning

Temporal reasoning permeates the task of interpreting the earth's history and structure. Geologic rhetoric is temporal. Many of the important concepts in geology encapsulate time: its key words imply stories. Consider "eruption" and "erosion" for example. Eruption and erosion happen as a sequence of events; each of these concepts refers to a process that unfolds in time.

An interpretation cannot be true if the events interpreted do not cohere in time—in sequence and in synchrony. Time referees among geologic explanations every time.

The jumbled rocks of the Hell Creek Formation in North Dakota hold a graveyard of creatures, great and small, young and old, encased in the debris left by a massive seiche wave. Violent shaking of the earth set an ancient, inland sea near Hell Creek in motion with catastrophic effect. This episode of dinosaurian death came in a single instant on a single day. Was the culprit a distant impact of an extraterrestrial object? If the dating of local burial and distant cratering agrees, then probably so—especially if, a short time after the creatures perished, glass beads rained down to further entomb the corpses in the rubble. They did.[10]

Comparative Thinking

Good geologic reasoning depends upon making comparisons in terms of carefully constrained, yet ultimately (and even necessarily) ambiguous, categories. The reason is simple: every feature of the earth has a history; features may share elements of their histories, but none have exactly the same history. The members of a geologic category are unique—the same as members of a family or a collection of snowflakes. Mountains, faults, crustal plates, fold belts, basins, and many more categories of the objects of interest in geology boast unique entries.

All river deltas are similar to each other and at the same time different from one another. In contrast, individual atoms of carbon 12 are utterly indistinguishable and completely interchangeable—a hallmark of conceptualization in chemistry. Deltaic sediments typically grade onto the seafloor into fan-shaped structures. Comparisons among deltas and between deltas and seafloor fans yield insights into the processes that formed them. History makes them different and inferring this history adds up to become an explanation. A carbon atom, on the other hand, does not come wrapped in the history of its past interactions; the Grand Canyon of the Colorado River does.[11]

Geologic reasoning responds to challenges of scale and the contingent nature of historical phenomena. Often analogical approaches—comparison and contrast—play an important role in geologic inquiry. For example, how an ancestral, pre-present-day major river draining the Colorado Plateau may have carved its way to the ocean has fueled contentious debate for many years. Perhaps for a time the river simply faded away without reaching the ocean. Utah geologist Joel Pedersen has proposed that the Colorado infiltrated Paleozoic limestones beneath its riverbed, dissolved masses of carbonate rock, and emerged to the west in desert springs only to evaporate and redeposit volumes of limestone. An analogue exists in the present: the termination of rivers in China's Tarim Basin and Angola's Kalahari Desert.[12] The analogue makes tentative interpretation credible—not certain, but persuasive.

Appreciating Scale

Understanding "scale" is also crucial to geology as a way of knowing. Dynamic interactions encompassing both physical and living systems span the vastness of geologic time and occur from micro- to macroscales in time and place. The barrier of "scale" challenged geologic explanation until, in 1830, Charles Lyell demonstrated how "causes now in operation" could account for events that occurred in the remote past. Although appreciation of time's vastness may remain a psychological obstacle to novice learners, grasp of geologic time is a rewarding pleasure once learned meaningfully.[13]

Story Form

The meaning of a sentence depends upon the meaning of its words. In turn, the precise meaning of the words depends upon their context in a sentence. This process of interpretation iterates from one level of text to another: sentence, paragraph, story, and sociohistorical context.

The earth is a text to interpret in an analogous fashion. In "earth stories," philosopher Robert Frodeman's term for geologic explanations, context gives meaning to the parts and parts give meaning to the whole.[14] A change in either may prompt reinterpretation. For example, the idea of crustal plates suggested new interpretations of prior facts about continental margins. New facts about plate boundaries catalyzed revision about crustal structures and processes.

In geology, minerals in the rock signify a process; the exposure of an outcrop is their "paragraph" and the landscape a "story" to be told—one that may well lead to reinterpretation of the outcrop. The imposing walls of volcanic tuff in Oregon's Smith Rocks State Park bake in the sunshine of the High Desert. By comparison with nearby beds of volcanic ash and analogy to the present-day range of Cascade volcanoes, these tuffs were interpreted for several decades as the lithified ashfall from arc volcanoes paralleling a subduction zone trench. Presumably, the eruptive centers lay buried.

A stunning new interpretation of the regional geology of the Crooked River Basin has overturned this interpretation. Rather than the products of an arc of subduction zone volcanoes, the tuffs of Smith Rock signified caldera style eruptions from forty-five million years ago.[15] The overarching story of collision between the North American and the Pacific tectonic plates took a dramatic twist. Convincing evidence that a mantle plume erupted through the crust led to the reinterpretation of the cliffs at Smith Rocks as fragments of a vast crater. Many of the facts remained the same, but the story had changed, the story of a hot spot tracing across North America to Yellowstone country.

CODA

Lists of objectives and matrices of topics, the standard fare of curriculum design, too often fail to cohere, thus obscuring how ideas derive meaning. They break the correspondence responsive to context between events of interest, their disciplined representations, and explanatory ideals. Geologic thought strives to uncover the causes that have transformed the earth over eons. Temporal facts referee among competing claims; scale presents obstacles. Stories—especially explanatory ones—set the conceptual boundaries of meaning. Analogues and comparisons enter and exit stage on cue.

Metaphorical use of concepts creates opportunities for novel thinking to capture aspects of reality otherwise ignored. In a fundamental sense, metaphor anchors representation. Maps, cross sections, block diagrams, and other visual forms mediate thought and reality when the problem is one of solving geologic problems.

Ideas cohere in context. In isolation, coherence fails and learning flounders; its primary recourse becomes rote memorization. Incoherent thoughts mark the mind of a rote learner attempting to solve difficult problems. The confusion that results feels painful. Knowledge exists as relationships; its structure gives it vitality. Well understood, knowledge transforms the meaning of experience.

NOTES

1. René Magritte, *The Treachery of Images (This is Not a Pipe) (La Tarion des Images [Ceci Ne's Pas Une Pipe])*, 1929, Los Angeles County Museum of Art (LACMA), Los Angeles, accessed August 15, 2020, https://collections.lacma.org/node/239578.

2. Alfred North Whitehead, *The Aims of Education and Other Essays* (New York: The Free Press, 1929), 2, 12.

3. John Dewey, *The School and Society and the Child and the Curriculum* (Chicago: University of Chicago Press, [1902] 1990), 200.

4. Timothy F. H. Allen and Thomas W. Hoekstra, *Toward a Unified Ecology* (New York: Columbia University Press, 1992), 75.

5. Robert Fillmore, *Geological Evolution of the Colorado Plateau of Eastern Utah and Western Colorado* (Salt Lake City, UT: University of Utah Press, 2011).

6. Mike Searle, *Colliding Continents: A Geological Exploration of the Himalaya, Karakorum, & Tibet* (Oxford: Oxford University Press, 2013).

7. Philosophers and historians of geological science agree on this point. See Robert Frodeman, *Geo-Logic: Breaking Ground between Philosophy and the Earth Sciences* (Albany, NY: State University of New York Press, 2003), Martin J. S. Rudwick, *Earth's Deep History: How It Was Discovered and Why It Matters* (Chicago: University of Chicago Press, 2014), and David B. Kitts, *The Structure of Geology* (Dallas, TX: Southern Methodist University Press, 1977).

8. James Lawrence Powell, *Night Comes to the Cretaceous: Dinosaur Extinction and the Transformation of Modern Geology* (New York: W. H. Freeman and Co., 1998). Powell provides an excellent narrative of the story of end of Cretaceous dinosaur extinction caused by the impact of an extraterrestrial object—often referred to as the "Alvarez hypothesis" after Louis and Walter Alvarez's discovery of the iridium anomaly at Gubbio, Italy.

9. Carol E. Cleland, "Common Cause Explanation and the Search for a Smoking Gun," in *Rethinking the Fabric of Geology*, Geological Society of America (GSA) Special Paper 502, ed. Victor R. Baker (Boulder, CO: GSA, September 1, 2013), https://doi.org/10.1130/2013.2502(01).

10. Robert Sanders, "66 Million-Year-Old Deathbed: Linked to Dinosaur-Killing Meteor," *Berkeley News* (March 29, 2019). Sanders summarizes research on the effects of the Cretaceous impact at Chicxulub, Mexico, on creatures living in what is now Montana as discovered in the Hell Creek Formation, Tanis site. The original study is Robert A. DePalma, Jan Smit, David A. Burnham, Klaudia Kuiper, Phillip L. Manning, Anton Oleinik, Peter Larson, Florentin J. Maurrasse, Johan Vellekoop, Mark A. Richards, Loren Gurche, and Walter Alvarez. "A Seismically Induced Onshore Surge Deposit at the KPg boundary, North Dakota," *Proceedings of the National Academy of Sciences (PNAS)* 116, no. 17 (April 23, 2019): 8190–8199; accessed August 15, 2020, https://doi.org/10.1073/pnas.1817407116.

11. Charles R. Ault Jr., "Criteria of Excellence for Geological Inquiry: The Necessity of Ambiguity," *Journal of Research in Science Teaching* 35, no. 2 (1998): 189–212.

12. Joel L. Pederson, "The Mystery of the Pre-Grand Canyon Colorado River—Results from the Muddy Creek Formation," *GSA Today* 18, no. 3 (March 2008): 4–10.

13. David Hawkins, "Critical-Barriers," *Outlook* 30, no. 4 (Boulder, CO: Mountain View Publishing co., 1978): 16–24.

14. Friedman, *Geo-Logic*.

15. Jason D. McClaughry, Mark L. Ferns, Caroline L. Gordon, and Karyn A. Partridge, "Field Trip Guide to the Oligocene Crooked River Caldera: Central Oregon's Supervolcano, Crook, Deschutes, and Jefferson Counties, Oregon," *Oregon Geology* 69, no. 1 (2009): 25–44.

Part V

COMMUNITY PURPOSE

Chapter 23

Simple Automata

The ants appeared at school with the arrival of spring weather, happy to move into the first graders' ant farm. Using modest magnification, the six- and seven-year-old children observed them with care. "All of their legs are on the thorax. The queen has wings," explained Kelly in her most serious tone of voice.

Kelly proceeded to cut out the drawings of ant body parts and arrange them properly connected. A while later, the finished product was ready to perform: an ant marionette, head and abdomen suspended from one chopstick; the thorax, with legs attached, strung to another.[1] To make an ant marionette is to design and engineer an appealing toy symbolizing a child's knowledge of insect anatomy.

Good things happen when nature, technology, and society go to school. Creative projects unfold. Mutual respect develops. Celebrations displace testing. Teachers take risks. Students learn that they matter and that the subjects they are learning matter. Feelings of attachment to place and membership in community flourish. The merger of engineering, puppet performances, and Nature Study through project work, believed John Dewey, makes school "a genuine form of active community life, instead of a place set apart in which to learn lessons."[2]

THE AUTOMATA OF ARBOR

Down the hall, fourth-grade children turn their attention to a small stage in anticipation of the drama about to unfold. A crank turns and the curtains open. Pugilism ensues. Two men stand facing each other, their arms pumping back and forth as they dodge each other's blows. The action repeats . . . and repeats with each turn of the crank. "How do you make them do that?" asks Roberto.

The answer to Robert's questions lies beneath the stage. The open curtains reveal not only the actors in action but also the mechanism below the stage controlling their movements. Gears couple the crank to two spools joined by a belt made from a rubber band. A peg near the rim of each spool sits in a slot formed by two strips of wood. The wood strips extend to grip and rub a vertical pole. The pole spins and, thanks to some clever gearing and joint construction hidden by costumes of the fighting figures, the arms box and joust. Mechanical expression of an elegant design delightfully tells a snippet of story. At this school, puppetry has evolved from strings to gears.

The scene depicts Robin Hood's legendary encounter with the towering Little John met while crossing a log over a stream. Due to their good-natured stubbornness, neither will yield passage and they begin to tussle. Robin wields his oak staff to no avail—with a powerful thwack, Little John knocks him off the log and into the creek. That's how the story goes; on the automaton stage there's no end to the struggle. The crank just keeps turning, and Robin Hood never loses his footing.

Celebrate Everything

The child who engineered this simple machine toy smiles while receiving a round of applause and the celebration continues. Each member of the class, in turn, presents a finished automaton well stocked with cams, gears, pulleys, and levers. Sharing marks the end of nearly ten weeks of project work. The automaton synthesizes the children's understandings of multiple topics that reach back to the start of the year at Arbor School of Arts and Sciences.[3] The purpose of the automaton project is to "celebrate everything," explains the teacher of the Intermediates (blended fourth and fifth grades).

What does "everything" mean? Arbor structures curriculum with themes that alternate in two- or three-year cycles. For the Intermediates, the themes are "Environments" followed by "Inventions and Discoveries." Both integrate the sciences and humanities through projects big and small accomplished by creative coordination of instruction from one subject to another—mathematics and art included. "Inventions and Discoveries" culminates in the design and engineering of an automaton. "Environments" closes with elaborate mapmaking that treats data as art and art as data, including time lines from the personal to the cosmological. Finished products represent the meaning of the year's theme on a personal level—a tangible expression of a part symbolic of the whole.

In one mapmaking project, a carved pumpkin served as a lampshade. The projection of the face onto a wall demonstrated the distortion of mapping the surface of a sphere onto a plane. This "Mercator projection" turned the pumpkin's stem into a long, horizontal line and enlarged the eyes relative to the nose, similar to how such a projection exaggerates the area of Greenland.[4]

Arbor has invested thirty years of effort and reflection in its thematic approach. Emphases on historical epochs change over time as cultural evolution unfolds; history looks different in the changing light of the present. There is, nonetheless, constancy: the curriculum values the history of human ideas embodied as inventions or discoveries. There are many to choose from, of course. Some inventions, such as writing, are so fundamental that their prospect of being deleted during an episode of revision is vanishingly remote.

Arbor School children author stories, set type, print, and bind books of their own. Now that the school has entered its fourth decade, there are parents who attended Arbor whose children are second-generation bookbinders.

"Inventions and Discoveries" encompasses the technical as well as the conceptual: the principle of the arch and the development of algebra, for example. Arbor's Intermediates learn the succession of astronomical views of the world as taught by ancient civilizations and as developed by Galileo during the Renaissance. From time to time an enterprising automaton engineer may select the story of a constellation—a hunter, a bull, an eagle—to turn into a toy operated by a crank.

Archimedes' Vending Machine

The "Inventions and Discoveries" theme has a decidedly Western Civilization cast; fall begins with a narrative of ancient Greece. The year winds down with the narrative coursing through the life of the mind in the centuries of the

European Renaissance. Early on, the teacher introduces the class to the invention of Archimedes' screw. This clever design wraps a ramp around a cylinder to solve the problem of moving water for irrigation uphill. In response to learning the history of this invention, children gather in science class to construct Archimedes' screws with polyvinyl chloride pipe.

The science lesson focuses on changing variables and the need to manipulate just one at a time while keeping the others constant. Rotating a coiled wire within the polyvinyl chloride tube moves material up an inclined plane (the bottom of the tube). In their experiments testing variation in design, children change tilt, coils count, tube diameter, material, and rotation speed.

Ramps make lifting easier, and coiled ones have many uses. They work to overcome resistance, as do all simple machines, and that's where science class in a year devoted to the study of invention spends much of its time. Intermediates study levers, pulleys, and wedges with examples of simple hand tools: wrench, pliers, scissors, hammer.

The teacher shares a picture of a vending machine. Pressing a button rotates a coil to release the purchase—an Archimedes' screw for lowering rather than lifting. Students proceed to search the classroom, playground, and schoolyard for examples of simple machines. Bicycles abound with examples: handlebar, wheels and axles, gears, brake levers, cranks.

Simple Machine Basics

A broom is a lever that increases speed—sweeping is a lever in use. One hand functions as the fulcrum; the other hand does the work. The broom pivots at the fulcrum as the long arm moves widely and rapidly. It sweeps. The hand applying effort moves back and forth through a shorter distance. A broom in use is roughly analogous to a teeter-totter oriented vertically.[5]

Friction inevitably dissipates some of the energy put to work by a simple machine. In the design of automata, friction can be a help or a hindrance. As an end-of-the-year project, the design of each automaton revisits and reinforces where the year in science began: the study of force and simple machines.

At midpoint in the design phase of the automaton project the children confront a challenge presented by the teacher: "See if you can design an apparatus to make something jump up and down while sliding forward." The challenges call upon them to draw from their knowledge of simple machines to build a system of interactions. The class divides into small teams that work diligently in order to be ready to propose a solution verbally and with sketches.

The class reassembles to listen to the solutions and critique the drawings presented by each team. An asymmetrical cam pierced by a rotating rod (a cam is a shape cut from a cylinder—picture a rotating cookie) and pushing against a piston can make a figure rise; a sharp bump on a cam causes quick-lift action—a jump.

The teacher adds a wrinkle to make the problem more difficult: make the figure jump twice, then pause, then jump twice again all while sliding forward. Solutions rely upon the design of the cam—where to place and how to space bumps and bulges. A rod may have multiple cams of different shapes making pistons rise and fall in or out of synchrony. Sliding forward is easy: just push the entire assembly along an axle.

Cams make movements start and stop. Levers increase speed. Gears control rates. The class is ready to design and engineer automata to express their understanding of simple machines in the context of synthesizing a topic studied as part of the year's theme.

PROJECT PROPOSALS

Deciding what part of the year to celebrate in the design of an automaton is personal. Children may freely choose what to celebrate, but their proposal must link to the year's theme and demonstrate good grasp of the topic. Most anything is fair game if somehow expressive of classical civilization's or the Renaissance's discoveries and inventions. Recreating the balcony scene from Shakespeare's *Romeo and Juliet* might do, or perhaps the sword fight between Romeo and Tybalt.

Arts and Sciences

A proposal might celebrate mythical stories found among the stars, a song from music class representative of a historical epoch, or a biography read during the course of the year. Almost inevitably, someone feels inspired to construct an automaton representative of a discovery by Galileo Galilei or an invention of Leonardo da Vinci. In English class, da Vinci receives attention during a unit on biography, while in social studies (history), for his intellectual contributions to a scientific mindset.

Learning to use grids to make drawings to precise scales and gear ratios to change power or speed while coupling movements integrates mathematics with the design of automata, especially in the planning stage. Plans must be drawn to scale with orthographic (flat) views from the sides and top. To draw shapes in three dimensions, children use graph paper covered in triangles, not squares. The triangles are equilateral. The first exercise with drawing on triangular graph paper is to make cubes of different sizes, then rectangular prisms. Each drawing shows three sides viewed, in effect, from a corner point. Rounded shapes can be sketched, triangle by triangle, within the prismatic ones.

Proposals are critiqued by teachers and peers until settled upon. Plans are written and elements of design carefully drawn before moving on to construction. Each finished project celebrates a year of learning, integrating, and using knowledge in a unique fashion.

For ten weeks children have collaborated and cooperated in a shared intellectual journey. They have evaluated their own and each other's work. The teacher has challenged their imaginations, scaffolded their skills, modeled expected outcomes, and critiqued their work. Across subject fields, teachers have coordinated the curriculum in keeping with the theme—and drawn upon decades of experience to do so. The beginning of the year—not to mention all of the in-between moments—is present at the end.

Pull Toy Ornaments

Examples from years past stimulate the imagination while setting realistic expectations of what is possible. The first example is a pull toy, familiar to many children either as a gizmo to hang on the wall or an ornament on a Christmas tree. Pulling down on a string attached to hidden levers moves arms and legs up. Releasing the string lets them fall. The figure may be a clown or a bear—or most anything.

This familiar, simple toy focuses attention on joints and their movements—lever action. The pull toy stands at one end of a spectrum of design and engineering difficulty. The stage with Robin Hood and Little John striking at each other occupies the complex end of the spectrum.

In another creation, a bull charges a hunter. The bull-and-hunter scene personifies the remarkable constellation of Orion, the hunter, facing a charging bull, the constellation Taurus. These groupings of stars have been mythologized in cultures across the globe for thousands, perhaps tens of thousands, of years. They are easily observable during the school year in the night sky of fall and winter.

Several action figures stand in the middle of the spectrum of automata complexity. A ballerina twirls. A bear rises on its hind legs and lifts its paws. A gorilla jumps up and down. They are silly things with serious purpose: toys signifying knowledge.

LETTERS TO ARISTOTLE

At the end of "Inventions and Discoveries" and after the celebration of automata, each child composes a "Letter to Aristotle." In these letters they share examples of inventions and discoveries that have changed the world most significantly since his era. They wonder which changes might most astound him. Each letter expresses the writer's synthesis of a year of study.

Community

The thematic, interdisciplinary approach at Arbor offers students extraordinary opportunities for growth. Project work abounds. Celebrations that consolidate in-depth study, such as the performances of the automata, demonstrate what a community of students and teachers can accomplish together.

The challenge to balance topic integration with developing skills within subject fields never ends. Arbor's curriculum acknowledges the value of disciplinary integrity in the arts and sciences as a context for exploration and investigation as well as the need to master particular skills in an orderly progression. At the same time, the curriculum values freedom for both students and teachers to make choices. Every aspect of the school recognizes that curiosity and freedom, fundamental values of the arts and sciences, are essential to the development of creativity.

Sharing

Arbor prizes spectacular celebrations, presentations, displays, and work products. No practice is more central to the conduct of schooling at Arbor than sharing. Students articulate the "how" and "why" of what they are doing. The arts are ever present. Artistic expression attaches emotion to content to share not only understanding but also the feeling of its significance.

Arbor is an independent and modestly expensive school. The administration strives to keep per pupil expenditures in the range of well-funded public-school districts, but the school must depend on tuition income to survive. Not everyone has the privilege of attending such a school. But every child deserves a school devoted to creating a caring community. Arbor demonstrates for everyone the centrality of creating, thinking, sharing, and collaborating to the accomplishment of this aim. Love and respect for learning walks hand in hand with love and respect for each other, especially when both are celebrated. Repeatedly.

To write, one must write. To create, one must create. To become a community, members must express respect, honor each other's talents and achievements, and engage in common purpose. Arbor students create from the earliest age, and the creations are authentic. They work with common purpose, share often, and never pass by an opportunity to celebrate each other and their work. The heart of the school beats to the rhythm of community building.

Creating community is a vital lesson for elementary school children to learn and for all citizens, whatever their age, to value. Community building depends upon collaboration, cooperation, mutual respect, celebration, artistic expression, reliable knowledge, equitable power-sharing, and open communication. When confronted with social and technical challenges, communities must rely upon these virtues. Teaching science beyond standards is a proper place to put them into practice.

NOTES

1. Susan Dunn and Rob Larson, *Design Technology: Children's Engineering* (New York: The Falmer Press, Taylor & Francis Group, 1990), 136–138.
2. John Dewey, *The School and Society and the Child and the Curriculum* (Chicago, IL: University of Chicago Press, 1990; original editions published in 1900 and 1902, respectively), 13.
3. Arbor School of Arts and Sciences, "Year 2: Inventions and Discoveries," accessed August 12, 2020, http://www.arborschool.org/intermediate-4-5.

4. Charles R. Ault Jr., "The Great Pumpkin Curriculum," *Science and Children* 22, no. 2 (October 1984): 13–16, accessed October 13, 2020, https://www.jstor.org/stable/43171346.

5. Mary Budd Rowe, *Teaching Science as Continuous Inquiry: A Basic* (New York: McGraw-Hill, 1978), 228.

Chapter 24

Egg Balloons

On a clear, calm day a few wispy cirrus clouds shaped like the runners of a sled drift high in the sky. Rural, upstate New York middle schoolers from Seneca County gather in front of their building to welcome the Weather Truck, a medium duty Ram 5500 outfitted with a workstation for running weather modeling software; two large, foldout touch screens; a canister of compressed "lifting gas"; hardware for balloon launching and tracking; and twenty virtual reality Oculus Rift goggles equipped for "vertical tours" through the atmosphere.

Eighth grader Sam turns the valve and the two-meter diameter, biodegradable latex, high-altitude weather balloon begins to fill with hydrogen. Liftoff happens moments later and immediately tracking data lights up the screen on the truck's foldout, touch screen monitor. The balloon's instrumentation is a marvel of low-cost sensors—technology that puts authentic science in the hands of young citizens.

Affordable, high-altitude ballooning has captured the imagination of middle schoolers for nearly a decade. In 2012, Bobble-head Barack Obama made its stratosphere debut, followed by twelve-year-old Lauren's Hello Kitty doll in 2013, as reported by ABC News and viewed on YouTube.[1] The viral success of Lauren's Hello Kitty video indicated the stratospheric attraction of ballooning to young adolescent learners. Her launch was a science fair entry.

PROJECT BOOTLEG BALLOONS

Bootleg Balloons, a novel, community-engaged science program created by Cornell University in collaboration with the Paleontological Research Institution/Museum of the Earth and Wicked Device, LLC, sponsors the Weather Truck. The program provides opportunities for young adolescents from rural families to collect data for improving regional weather forecasts.[2] To launch a weather balloon equipped with affordable instruments for recording atmospheric data is to place atmospheric sensing and climate modeling within reach of citizens and schools.[3]

Donning the virtual reality goggles simulates the experience of riding along with the balloon. The visual grandeur and the augmentation of reality combine to make the invisible visible. The virtual journey provokes wonder. Gauges indicate an airspeed with respect to the ground of well over 100 miles per hour. Children listen for the sound of the high winds. They hear nothing. The balloon moves with the wind; there are no leaves to rustle, buildings to stream around, ears to brush. The intense energy of the atmosphere is silent, surprising everyone.

Wicked Eggs in a Basket

"Be careful when putting all of the eggs in the 'basket,'" says the teacher. The eggs, engineered by Wicked Device, are a set of sensors and instruments for measuring atmospheric conditions.[4] The "basket" (a small gondola) is the package of "eggs" suspended meters below the balloon. Eggs measure properties of the atmosphere as the balloon ascends. The devices do, in fact, look like eggs. A Global Positioning System instrument tracks the balloon's movement—rate of rise, speed over the ground, and location (figure 24.1).

The Air Quality Egg System (AQES) stores data and generates interactive visual output tailored for classroom use (thanks to previous NSF funding). A digital camera records high-resolution images of clouds throughout the balloon's journey to the stratosphere where it bursts.

The newest egg measures soil moisture from aloft—quite a trick, but data much-prized by the farming and aviation communities. Soil moisture is of obvious importance to farming because seasonal trends in soil moisture as climate changes have profound implications for regional agriculture. Soil moisture is also a variable that influences low-level fog formation, information

Figure 24.1 Sixth-grader Elinor's Science Journal Sketch of the Weather Balloon.
Source: Photograph by Toby Ault. Used with permission.

required for calculating air density at low elevations. This information is critical for guiding aircraft.[5]

As the balloon ascends, sensors begin to measure temperature, pressure, humidity, carbon dioxide concentration, windspeed, and particulate matter in the one-micron range (the size that can enter the human bloodstream). Detecting ozone levels in the stratosphere is forthcoming.

While the balloon flies the class takes turns with the virtual reality goggles. The goggles place them in flight on a balloon high above the local clouds to the stratosphere, the images captured during previous ascents equipped with a GoPro Hero4 Black camera with a resolution of 1,920 × 1,440 pixels capable of shooting at forty-eight frames per second. Drifting at twenty miles high feels like being at the edge of space. All the weather they have ever known lies in the troposphere ten miles below. At the apex of their ascent, the dark sky surrounds a brilliant, white sun, a star in close-up. Against the horizon, a thin, blue band rings the earth.

Egg data is destined for downloading to an online database. Later in the year, as other schools enter their data, students will be able to inspect, for example, air quality data for the size and prevalence of aerosols from smoke or allergens from pollen over the region.

Carbon dioxide levels change with the seasons and altitude; cold, high pressure cells trap this gas closer to the surface. As the database grows, so does its potential to answer questions about trends and cycles of carbon dioxide concentrations in the atmosphere at different altitudes and during different seasons.

Dairy Farm Data

High in the stratosphere temperatures drop to dozens of degrees below freezing. The rubber balloon, its girth expanded nearly to ten meters, meets a chilly end. Sudden rupture rips it into shreds. A parachute lowers the scientific cargo to earth seventy-five miles downwind from the launch site.[6] The onboard global positioning system tracks its location within a radius of 100 feet.

A van of meteorology students from Cornell University navigates country roads to find where the weather balloon has landed. En route to fetch the basket of eggs, they determine air density at different altitudes and calculate the force accelerating the balloon upward. The Amateur Ham Radio Club of Cornell picks up pings broadcast from the balloon. A phone call to the club lets the van driver know exactly where to go. After driving for two hours, the chase team finds themselves at the entry lane to a dairy farm. The package of instruments they hope to recover lies in a cow pasture.

"Trump 2020" and "MAGA" signs along the road give the decidedly liberal Cornell professor and his progressive-minded students pause. They drive up to the farmhouse and knock on the door. A middle-aged dairy farmer, surprised to learn that a packet of atmospheric data has landed in his pasture, greets them with a pleasant smile. He listens to their description of research and request to recover the equipment, pleased that the instrument landing has harmed none of his cows.

A surprisingly high level of interest in the project by the dairy farmer follows the initial moments of unease. Though a skeptic of human-induced climate change, he has an avid concern for weather—especially patterns of rainfall and spells of extreme temperature. Excitement builds upon retrieval of the camera that rendezvoused with the stratosphere. Images of cloud tops towering over his pasture fascinate the dairy farmer, the same as everyone else.[7] The professor and the farmer bond over their shared hopes for accurate forecasting—one small step for political civility, one giant leap for inexpensive high-altitude ballooning.

Community Science

Twelve hundred dollars, three pilot launches, and three successful instrument retrievals later (all in dairy country), the Cornellians knew they had a feasible program at hand. The ancillary social benefits—at home and in the field—loomed as large as the promise of gathering atmospheric data widely and at low cost. Encounters in the field with members of the farming community brought realization of the value of linking school and university science to family interaction and community participation.

Solving problems at a community level requires reliable knowledge, widely shared, with potential to guide responsible actions. Partnering among citizens—even the young ones—and experts is crucial to finding workable solutions, especially when the problems are wickedly challenging. The availability of affordable, promising technology—eggs from the Wicked Device Company, for example—helps, too.

For adolescents, being perceived as a member of a science-oriented community enhances identity. The recognition of their scientific competency at home and in the community creates virtuous cycling among interest, knowledge, confidence, stature, and action.[8]

THE RHYTHM OF THE SEASONS

Students in rural New York participating in the Bootleg Balloons program take home soil moisture, air quality, and carbon dioxide sensors. They carry out investigations throughout the year with these "eggs." Monitoring soil moisture from space is feasible, but there are no affordable eggs for doing so from high-altitude balloons. Carbon dioxide sensors can go most anywhere, as can devices that detect minute particles of pollutants.

With these sensors, plus traditional rain gauges and barometers, students monitor rainfall, soil moisture, and change in air pressure as weather systems move through their neighborhoods. Surface runoff quickly raises the level of local streams; soil moisture decays exponentially. The task becomes comparing recordings made on the ground with what the weather is doing above.[9]

Moisture in the Soil

Plants modify soil moisture at the depth of a meter and below, and, therefore, soil moisture behaves differently at different depths. Using a coring tool, students dig to depths of ten centimeters, thirty centimeters, and one meter to place sensors. The "surface" may freeze to a depth of ten centimeters and vary dramatically in average moisture levels on a seasonal scale. Spring arrives, buds burst open, and leaves begin to grow and function. Leaves pump water into the atmosphere. Roots pull water from the ground, changing water pressure at depth, and may reduce the rate of groundwater recharge of springs despite high rainfall.

There are diurnal, seasonal, and annual cycles of moisture, carbon dioxide concentration, and humidity to detect in both ground and air. Studying these processes at home and school makes invisible, backyard, microclimate phenomena visible. When scaled regionally and decadally, these data reflect climate changes affecting the frequency and intensity of drought and precipitation.

The Egg in the Maple Tree

During a week in March, sixth grader Elinor and her younger brother, Axel, conduct a simple experiment to gain familiarity with the sensors and the data they generate. They suspend air quality monitoring eggs to measure temperature, air pressure, relative humidity, and carbon dioxide levels in a backyard silver maple tree at a height of fifteen feet. The eggs in the maple tree serve as experimental controls. Elinor places a second set of sensors in a fish tank inverted over a bed of rapidly growing bean plants.

Most days during the study are clear but cool with a couple of nighttime temperatures dropping below freezing. The greenhouse effect raises the temperature within the tank on sunny days and keeps humidity elevated with respect to the outdoor environment of the maple tree. Air pressure within and outside the tank does not differ (see figure 24.2).

Plants draw in carbon dioxide during the day as they photosynthesize and release oxygen. Plants require energy twenty-four hours a day and so must

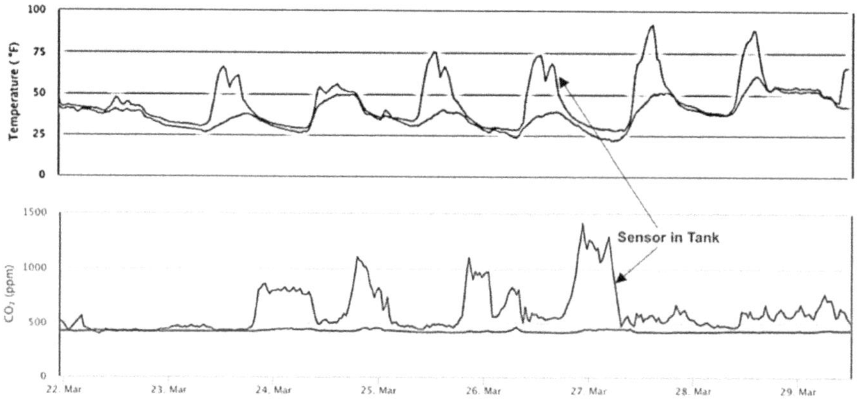

Figure 24.2 Measurements of temperature in degrees Fahrenheit (top) and carbon dioxide in parts per million (bottom) taken approximately every six minutes by two air quality "eggs" (AQES). Data from within the mini-greenhouse (an inverted fish tank) are higher than the control placed in a backyard silver maple tree. Temperature peaks in the midafternoon; carbon dioxide, at night. The upper, highly peaked lines record data from the sensors ("eggs") in the tank. *Source*: Elsa and Isak Leijonhuvud Ault. Used with permission.

respire constantly. Carbon dioxide and water vapor are the end products of respiration. At night, in the absence of photosynthesis, carbon dioxide levels rise. Carbon dioxide peaks after dark but before midnight because respiration slows in the early hours of the morning as temperatures reach their lowest. The tank sensors recorded dramatic rises and falls in carbon dioxide levels on a diurnal (daily) cycle. The experiment tunes investigators Elinor and Axel to the rhythms of atmospheric changes.

MAKING THE INVISIBLE VISIBLE

The problem of interpreting weather data is one of understanding atmospheric interactions on multiple spatial and timescales as masses of air jumble and tumble together around the earth, pressure gradients set winds in motion, winds steer their course, and the spin of the planet twists their direction. Forecasters seek to know what is happening in many places at the same moment in order to figure out what will likely happen at a specific place at a future moment. They must frequently update the observations, repeat calculations, and draw new maps.[10]

Much of what happens in the atmosphere remains invisible, clouds excepted. Even visualizing the airflow within a very visible tornado presents terrific challenges, not to mention the conditions that spawn tornadoes. The same holds true for regional and global weather phenomena from daily to decadal scales. On a local scale, a downburst of wind may topple trees on one side of the street while not disturbing those on the other. Ripples in the jet stream (peaks and troughs oscillating north and south) have nodes spaced on a continental scale. On a weather map, the sinuous jet stream resembles an immense fire hose stretched across North America from coast to coast. It's the region's weather delivery system.

Visualization Tools

Visualizing the state of the atmosphere and its future depend upon collecting a host of observations. Typically, the public encounters these observations assembled in visual form as a "synoptic weather map," a view of weather patterns and conditions on the scale of several hundred to thousands of miles as seen on television weather forecasts or in the daily newspaper. Other tools for displaying water vapor concentration, winds aloft, pressure gradients, and cloud cover, thanks to remote sensing and satellite imagery, are ubiquitous in weather reporting and forecasting.

A synoptic weather map represents ground level phenomena, but, of course, a weather front between two masses of air with contrasting characteristics (humidity, temperature) extends vertically through the troposphere. On the local scale, a tornado is embedded in the front where one air mass advances upon another. (The word "front" descends from military use in locating battle lines.) The large-scale synoptic pattern resides within the structure of global circulation and this structure changes seasonally—the largest scale changes reverberating back to the local scale of a front boundary at a place, at a time. To a first approximation, "climate change" refers to the regional effects of altering, on average, the global structure of circulation.

It's neither safe to be present in the midst of dangerous weather systems nor possible for people to ride a balloon to the stratosphere (though that situation may change[11]). Virtual display can make the hidden properties and motions of the invisible atmosphere visible. In addition to the scenic perspective provided by camera as a balloon rises to over 100,000 feet, programmers tint the sky to represent temperature, humidity, or pressure and add swarms of arrows ("streamlines") to show wind direction and speed.

Sense of Atmosphere

Understanding the data for forecasting daily weather and recognizing the challenges of making the invisible visible on multiple scales corresponds tightly with the demands of climate modeling. There is importance to teaching the impact of fossil fuel burning on the future of the climate, of course. Equally fundamental is to develop a realistic sense of the complexity of the atmosphere and a feel for how meteorologists and climate scientists are able to figure out what's happening where the sky meets the earth.

The goal of the Bootleg Balloons project is for young citizens to develop an intuitive feel for the structure and dynamism of the atmosphere. Everyone experiences weather directly. The atmospheric conditions, however, that must be known for the purpose of making forecasts exist on scales beyond the threshold of perception. Synoptic weather maps, contours of pressure at different altitudes, vertical cross sections, and global circulation diagrams render imperceptible atmospheric patterns to scales humans may interpret. They are tools used by experts for seeing through time and space. Virtual reality converts this information into simulated experiences that carry novices beyond the threshold of immediate perception into the realm of insight, to the achievement of a sense of atmosphere.

The investigation of backyard microclimates is open to all ages of people in all places. The thrill of ascending vicariously to the stratosphere has become accessible to many. Technological advances make tracking of trends and cycles in daily weather data related to climate change a feasible enterprise for citizen scientists. Opportunity for science education knocks loudly; promise abounds.

NOTES

1. Katie Kindelan, "Girl Blasts Hello Kitty Doll into Space," *ABC News* (February 5, 2013), accessed August 12, 2020, https://www.yahoo.com/gma/blogs/abc-blogs/girl-blasts-hello-kitty-doll-space-191135994--abc-news-tech.html.

2. Toby R. Ault, principal investigator, "Bootleg Weather Balloons to Activate STEM Identity in Rural Farming Communities," Engaged Curriculum Grant, Cornell University, accessed October 7, 2020, https://engaged.cornell.edu/recipient/bootleg-balloons/.

3. Ibid.

4. Dirk Swart/Wicked Device, LLC, "Small Business Innovative Research (SBIR) Phase II: A STEM Toolkit Enabling Global Air Quality Experiments," accessed August 12, 2020, https://www.sbir.gov/sbirsearch/detail/1500225.

5. Andrew Blum, *The Weather Machine: A Journey Inside the Forecast* (New York: Ecco/HarperCollins, 2019), 96.

6. Patrick Cullis, Chance Sterling, Emrys Hall, Allen Jordan, Bryan Johnson, and Russell Schnell, "Pop Goes the Balloon!: What Happens When a Weather Balloon Reaches 30,000 meters of Altitude?" *Bulletin of the American Meteorological Society* 98, no. 2 (February 2017), accessed August 12, 2020, doi.org/10.1175/BAMS-D-16-0094.1.

7. Ault, "Bootleg Weather Balloons."

8. Heidi B. Carlone and Angela Johnson, "Understanding the Science Experiences of Successful Women of Color: Science Identity as an Analytic Lens," *Journal of Research in Science Teaching* 44, no. 8 (2007):1187–1218, accessed August 12, 2020, doi:10.1002/tea.20237.

9. Funding of Bootleg Ballooning, following the pilot stage, presumed activities with participating schools would begin in spring, 2020, and continue through 2021. The Covid-19 pandemic has suspended these plans until 2022. The project, an initiative of Engaged Cornell, stands as an exemplar of selecting a science needed to contribute to community betterment and bears careful watching in the future.

10. Blum, *The Weather Machine*, 58.

11. Leonard David, "Balloon Flights Bring Near-Space Exploration to Masses," *Space* (March 7, 2013), accessed August 12, 2020, https://www.space.com/20089-near-space-balloons-science.html#:~:text=Weather%20balloons%20can%20rise%20to,the%20launch%20site%2C%20Maydell%20said.

Chapter 25

Competing Forecasts

The Panic of 1893, triggered by a run on gold reserves, created a severe, four-year depression in the United States. Frank Baum's classic, *The Wizard of Oz*, parodied the financial crisis and the populist response.[1] Dorothy and Toto, joined by the Tin Man, the Scarecrow, and the Cowardly Lion, venture on the yellow brick road (gold) to the Emerald City (the color of money), Oz (the unit for measuring gold). These characters represented women, labor, farmers, and African Americans (whose roar would be heard in the twentieth century).

Toto was a nickname for "teetotalers," the temperance movement folk. Wicked witches represented banking interests and monopolies on the east and west coasts, tormenting the land. The tornado is the cyclone of populism that swept Kansas in a time of financial panic.

RURAL DISTRESS

A recession preceding the panic reached the agricultural communities of New York in 1891. Anna Botsford Comstock, under the guidance of Liberty Hyde Bailey, recognized the need to respond to the distressed state of rural development and make rural life more satisfying and productive. In this context, Comstock introduced Nature Study lessons to children in rural schools. In addition, she designed professional development exercises for teachers to improve the quality of rural education.[2]

Cornell Professor of Meteorology, Wilford M. Wilson, authored the "Climate and Weather" section of Comstock's *Handbook of Nature Study*. His section addressed topics of water in the atmosphere and use of weather instruments in great detail. Wilson's entry was very didactic; at its conclusion, however, Comstock placed four enticing lessons. In the first, she directs a child to fill a "tumbler" (glass) with water and cover it with a piece of paper. Turned over, the water "mysteriously" remains trapped in the tumbler, held in place by air pressure. Teachers still conduct this demonstration.

In the second lesson, students construct a mercury barometer with mercury "an inch or more deep in the cup." Modern science teaching has banned mercury from school, of course, due to its toxicity. Comstock warned, "Be careful not to spill the mercury . . . or you will be as badly off as old Sisyphus with his rolling stone."

In the third lesson children learn to interpret weather maps and in the last lesson they explore phases of water—again, a completely modern set of elementary and middle school science activities.[3] The rationale for teaching these topics has only gained in strength. Weather and climate sit near the summit of humanity's concerns for the future.

WEATHER RESEARCH FORECASTS

Similar to a century ago, economic distress again troubles agricultural communities in central New York. In response, the Bootleg Weather Balloons project aims to engage schools in activities contributing to the productivity of their farming communities. Bootleg Ballooning, featuring schoolyard launches and backyard monitoring stations, serves middle schoolers (see chapter 24). In addition, the Paleontological Research Institution, the Cayuga Nature Center, and Cornell University have begun a project in climate model construction for rural secondary students.

Modular Modeling

On Cornell's Emergent Climate Research Laboratory's website, amateur model-makers can utilize modules for creating "Weather Research Forecasts." The modules incorporate parameters characterizing features of the land's surface and the flow of energy between the land and the atmosphere. A model's output forecasts wind, precipitation (and type), and temperature—as well as the likelihood of an extreme weather event—and the model is always in need of improvements, especially at the local scale.

Characteristics of land surface known to influence weather include "average roughness, soil moisture storage, and average reflectivity." Each of these averages may take on a range of values at some feasible scale of measurement in both time and space. Vegetation cover adds another set of parameters to consider: ground cover, urban surfaces, forest type, evaporation-transpiration rates.

Very, very many combinations of the values of these parameters are possible, each combination interacting in a unique way. The best combination for achieving the most accurate forecast is uncertain, so that means that the supply of models to test is, to put it mildly, "quite large." The coupling of land to atmosphere through the exchange of energy is indeed complex. Imagine ten values for each of just three land surface characteristics crossed with ten values for only three types of vegetation cover. That's $10 \times 10 \times 10 \times 10 \times 10 \times 10 = 1{,}000{,}000$ combinations!

Atmospheric data collected from ground level up to the stratosphere by the air quality monitoring sensors ("eggs") downloads to the Emergent Climate Risk Laboratory at Cornell. A public interface designed by the laboratory makes this database searchable by amateurs.[4] Each measurement captures the state of an atmospheric variable (temperature, humidity, etc.) at a time and in a place. Modelers input this data into the appropriate module. The modules, "knowing" these conditions, interact (by means of calculations) to generate forecasts. Their interactions approximate the changes forward in time expected from a given set of conditions.

Not all of the critical characteristics for forecasting are known at every scale imaginable and therefore the lab's Weather Research Forecasts come with uncertainty. A modeled forecast's resolution is typically on the order of one to five kilometers. Such a forecast struggles to resolve the fact that the sun may be shining in one neighborhood while rain falls intensely a few kilometers away.

The tools of modern meteorology follow in the tradition of the *Farmers' Almanac*. In the hands of computer-savvy high schoolers, modeling promises to increase the accuracy of local forecasts and help farmers plan for plowing, planting, and harvesting, as well as managing the timing of applying soil amendments.

The Gene in the Forecast

Consider the code for a climate model's module as analogous to a stretch of DNA, the molecule laden with a body's inherited information, in a chromosome. A change (mutation) in DNA—the genotype—causes a change in the body of an organism—the phenotype, often fatal, but sometimes advantageous. Forecasts are the "phenotypes of modeling" and coding is their DNA. Like DNA, changes in the code may interact in unintended ways—and behave as harmful mutations.

For example, an improvement in the module to account for convection in summer may interact with other, linked components of the module (and components of other modules) to degrade the forecast. Unlike Bilbo the Hobbit's ring in Middle Earth, "one model to rule them all" does not work well. A specific structure enables solutions while posing limits. Any model's structure prevents it from doing certain things correctly; all models have limitations.

There is, fortunately, an escape clause in the modeling contract: the errors in any one model typically fail to correlate with those of a completely independent model. An *ensemble* of many independently coded models thus has more strength than any one model. Therefore, when averaged together, independent models obtain more reliable solutions. The logic is simple: one goes left, one goes right, on average the path goes straight. The accuracy of prediction improves by eliminating uncorrelated error. Plow now? Spray later? The model average offers good advice to farmers.

Crowdsourced Modeling

The Emergent Climate Research Laboratory organizes its modeling tools in programming modules. One module simulates convection ferrying energy from land to atmosphere. Another module simulates the convective response of parcels of air to solar and infrared radiation. A third module treats the effects of cloud cover on radiation balance and heat exchange. Each module has a set of parameters that interact to simulate part of the process of convection. Together, the modules are capable of calculating millions of distinct forecasts. Which forecast corresponds best to reality?

The Weather Research Forecast software makes slight changes in parameters quite easy. One way to find a best—or at least better—combination in terms of accurately forecasting regional weather a few days in advance is to test crowdsourced model variants. One model, one forecast—a sort of "modelling democracy" emerges. Modelers may enter their forecasts in a competition because there is a clear standard by which to judge a forecast: the actual weather.

In this approach, each model slightly alters the parameterization. Thus, slightly different versions crafted by secondary school coders may run in competition with each other and, by trial and error, eventually even outperform National Weather Service forecasts.

Forecasting weather a week or ten days in advance follows the same basic consideration as climate modeling on much grander timescales. Climate, after all, is a region's average weather for a period of time.

Creating a Weather Research Forecast is a challenging task representative of climate science. The virtual experience of a balloon rise to the stratosphere communicates the nature of atmospheric phenomena. Both activities advance student perception of the nature of the atmospheric sciences. They demonstrate the demands characteristic of solving problems of weather forecasting and climate modeling—the nature of expertise in a purposeful context.

CULTIVATING PUBLIC TRUST

As so well understood by Dorothy in the Land of Oz, "There's no place like home." Climate change is impacting homes of farmers around the globe. Visualizing data and modeling weather in a virtual world places powerful meteorological tools in the hands of novices and empowers them to make a difference in their home communities. High school students learn to a run a weather simulation. They tweak the parameters and run the system again, each run visualizing weather a few days into the future.

The values of this activity are twofold. First, as local citizens, students contribute information to their community needed for making responsible decisions about farming. Second, as global citizens, they come to appreciate the nature of climate science and the role of modeling.

Cultivating public trust in climate science is a much-needed social value and understanding how science gets done in a particular context contributes to the establishment of such trust. The same as all other sciences, climate science responds with intelligence to the demands characteristic of its problems.

Listening to the silent sound of high-altitude winds, enjoying the beauty of the curvature of the earth against a blackened horizon, witnessing the leaves of the northern hemisphere breathe carbon dioxide in and out with the seasons, and "messing about" with the parameters of weather to improve local forecasting offer a promising foundation for placing trust in climate science.

Humanity has disturbed the near equilibrium state of the earth's climate system that has persisted for thousands of years. As a result, developing an intuitive sense of the system's being out of whack has become imperative. Making visible the weather processes playing out in backyards and scaling atmospheric phenomena beyond the threshold of perception into sensible forms are stepping-stones to placing trust in climate science.

Increasingly, communities hear climate forecasts that reduce to "Hot, with an increasing chance of megadrought and other extreme weather events." To think about weather, seasons, and climate is also to wonder what the roots of plants are doing—and hence the future of agriculture. The acceleration of climate disequilibrium threatens what humanity values: sustainable living, robust ecosystems, diverse flora and fauna, social and economic progress. Dysfunction plagues communities that fail to teach and to trust the knowledge required to overcome the consequences of a disequilibrating climate. As emphasized by philosopher Stephen Toulmin, the sciences pursued by a society respond to present circumstances, historical and social. School science is no different.

Picture a Caribbean farmer during a drought in the year 2035—a reasonably probable scenario, according to regional climate modeling in 2020.[5] Exceptional heat has depleted the local aquifers. Frequent blackouts occur because reservoir levels are too low to generate hydroelectric power. Now imagine the farmer lives in eastern Oregon or upstate New York or central Iowa or the Guanacaste Province of Costa Rica and has begun to experience persistent low levels of soil moisture. The cleverly phrased forecast, "hot with a chance of megadrought," has become reality.

NOTES

1. Quentin P. Taylor, "Money and Politics in the Land of Oz," *The Independent Review* 9, no. 3 (Winter 2005): 413–426.

2. Audrey B. Champagne and Leopold E. Klopfer, "Pioneers of Elementary-School Science: II. Anna Botsford Comstock," *Science Education* 63, no. 3 (1979): 299–322.

3. Anna Botsford Comstock, *Handbook of Nature Study* (Ithaca, NY: Comstock Associates/Cornell University Press, 1986), 780. Originally published in 1911.

4. Toby R. Ault, "ECRL: Emergent Climate Research Laboratory," Cornell University, accessed September 11, 2020, https://ecrl.eas.cornell.edu/.

5. Toby R. Ault, "On the Essentials of Drought in a Changing Climate," *Science* 368 (April 17, 2020): 256–260.

Chapter 26

Wicked Extinctions

The metaphor of "wicked problems," raised in the 1960s and 1970s, expresses the interaction of the "tame" problems of the natural sciences with the "wicked" ones of social science as they interact with each other and with public policy.[1] Wicked problems are difficult *and vital* to solve. Progress informs judgment as it reduces uncertainty. Wicked problems are addressed repeatedly and from multiple, disciplinary perspectives. They admit to no ultimate solution or definitive test—probabilistic reasoning dominates conclusions. They offer challenge and opportunity to education—the challenge to embed student learning in science that matters.[2]

THE WICKED PROBLEM OF WEATHER

The gold standard of weather forecasting is to extrapolate the future "big picture" at longer and longer timescales with more and more accuracy at increasingly local scales. Forecasts, once derived from mapping the weather over a widespread geographic area at a particular time, predict how conditions will change from place to place into the future. Adding knowledge of the conditions present through the vertical atmosphere (balloon data) enhances prediction immensely. Images and sensing from weather satellites further improved forecasts and their resolution.

Computing weather has always been a formidable challenge, but computational modeling has displaced making forecasts from geographic maps run in fast-forward mode.[3] Computational forecasting—climate represented as algorithm, weather as virtual reality—confronts the challenge of earning trust because actions based on projections have such great societal importance. Those who recoil from the implications tend to dismiss the science by cherry-picking its uncertainties, rather than acknowledging the reasonable constraints on uncertainty and the body of insight that has achieved consensus among practitioners. Earning trust is the wicked social dimension of climate science.

Weather forecasting on the scale of a week is now as reliable as the forty-eight-hour forecasts of 1975. By 2025, the prediction of severe weather events two weeks in advance will likely be achieved.[4] Regionally scaled modeling is more accurate than globally scaled modeling because the resolution of detail is, of course, much finer. How to respond to modeling's increasingly alarming and inevitably uncertain forecasts challenges political, economic, and social structures. Despite technological advances (and often because of them) weather forecasting and climate change modeling are truly complex, even "wicked" problems.

They are wicked not only because the input observations are astronomical in number and the interactions of variables complex but also because the need for and use of the predictions are vital to society—to public safety, travel and transportation, military activity, and, of course, agriculture. Daily living and all aspects of the economy are sensitive to the weather and expectations of what the weather will soon, or even not so soon, be.

An appreciation of climate science as a wicked problem underscores a primary purpose for learning science at school: to learn when to trust or doubt scientific claims with confidence as the case may deserve. Few individuals know how to compute climate change; a larger group holds the credentials necessary for evaluating the techniques of investigation and the conceptual understanding of climate researchers. Citizens need a sense of how the science gets done sufficient to avoid becoming overly credulous or cynical. People need to distrust snake oil science without becoming anti-science.

Conspiracy-mongering thrives on runaway suspicion. Being ignorant of relevant criteria for judging a science risks being duped by decontextualized statements. Appeal to the authority of "the scientific method" is a smokescreen that cannot resolve these risks and absolve an individual or a community of responsibility for responsible social action. For science teaching at all levels, warranting public acceptance of science—promulgating trust in reliable knowledge and reliable expertise—is a basic challenge.[5]

In the January 2020 edition of *The New Yorker*, Elizabeth Kolbert shared frightening figures. Just a couple of weeks earlier the average temperature across Australia reached 107.4 degrees Fahrenheit, culminating the warmest decade in centuries. Kolbert added that all-time temperature records were broken in France and Germany during 2019 and that during its summer a 100 people in the state of Bihar, India, perished from heat. She asked provocatively, "What will another decade of climate crisis bring?"[6]

The Covid-19 pandemic of 2020 displaced climate alarm with alacrity even as the march of habitat alteration and destruction induced by climate change accelerated the rate of species extinction.[7] The collapse of tourism changes the economic calculus of species preservation. The harm done to human health and the overburdening of health services by the pandemic promise to exacerbate species decline even further.

Organizing instruction in terms of wicked problems is not a recipe for mass indoctrination in doomsday scenarios. Yet climate problems and mass extinctions are real and teaching responsibility for solving them is an educational imperative. In Denmark, a school-museum partnership found an indirect path to raising concerns about climate change in the geologic record of a mass extinction event that occurred tens of millions of years ago.

THE WICKED PROBLEM OF EXTINCTION

The need to imagine possibilities in the future with reliable probabilities gives urgency to studying the past. As elusive as precise prediction in geoscience may be, earth history does inform the need for precautions and preparations. Earthquakes, extinctions, and climate change have exceedingly long histories. Through field study of mass extinction, secondary students can examine the wicked problem of climate change and learn how geologists use the past to understand the present and hypothesize the future. Danish educators familiar with a famous seaside outcrop of fossiliferous chalk and limestone have done so.

Stevns Klint

Abrupt changes in the fossil record mark the transitions from one geologic period to another. Five of these transitions have earned the status of "mass extinction events" and their causes are geologic puzzles with earthly tectonic, biologic, atmospheric, climatic, and extraterrestrial impact solutions. Even stellar causes, in addition to fluctuations in solar radiation, have received attention. Perhaps a periodic approach by "Nemesis," the name proposed for a possible companion star to the sun, disrupts life on earth. Of course, the coincidental convergence of several factors may ignite a flare-up in extinction rates.

Stevns Klint, a geological field site boasting inclusion in UNESCO's World Heritage List, rises forty-one meters above the coast of Zealand in Denmark. Extending for fifteen kilometers, the cliff exposes deposits spanning the boundary between the Cretaceous and Paleogene periods of earth history. Geologists date the boundary at sixty-six million years ago, the time when one-half of the species on earth became extinct due to the impact of an extraterrestrial object near Chicxulub, Mexico, on the Yucatán Peninsula.

Discovery of the crater and observations of sediment beds anomalously enriched in iridium (more than a hundredfold times natural concentrations) at locations scattered around the globe attest to the catastrophic collision that precipitated the Cretaceous mass extinction.[8] Ecosystems collapsed on a global scale and the event ushered in the Tertiary Era—the Age of Mammals.

Fossil horizons above and below the boundary outcrop at Stevns Klint sample what happened in what was once seafloor and is now Denmark. Layers of chalk below and limestone rock above sandwich the "fish clay unit" that marks the auspicious date of sixty-six million years ago on the geologic calendar. Fossil abundance collapses abruptly as the chalk meets the fish clay. Above the limestone—beds built by colonial bryozoans—rest ice age clays.

The strata entomb organisms that lived and perished at a crucial "moment" in earth history—several hundreds of species large and tiny, from extinct sharks to planktonic microfossils.[9] From the famed repository of fossils at Stevns Klint, paleontologists are able to reconstruct environments and communities of organisms on either side of the Cretaceous-Paleogene boundary as well as the sudden extinction event itself. A new world replaced the dinosaurian monsters of the Cretaceous and their brethren in the seas.

"A variety of hypotheses—some ingenious and some preposterous—have been proposed to explain mass extinctions" wrote the authors of *Principles of Paleontology*. Agents of change—oxygen concentration, temperature shift—may stem from a common cause but have different effects on different environments, whether on land or in the ocean.[10] To merit "mass" status, an extinction event must depopulate the earth of a high proportion of extant species—on the order of half or more in a short time (on the geologic scale "short" can mean several hundred thousands of years or even much longer).

The fossil record suggests that 70 percent of all species vanished 450 million years ago (give or take several million). The cause remains under debate. Fire or ice? The culprit may be volcanism or even its virtual opposite, glaciation. Approximately 100 million years later, the next transition—from the Age of Fishes (Devonian Period) to the Coal Age (Carboniferous Period)—sputtered for 20 million years before reaching a similar level of species depletion. Life on planet earth sunk to its most tenuous hold a quarter of a billion years ago—during the "Great Dying" at the end of Permian and start of Triassic time (the period of dinosaur origins). More than 90 percent of species of marine life exited the stage and more than two-thirds of life on land as well.

The Triassic Period yielded to the Jurassic 200 million years ago, again with an extinction signature in the range of 70 percent loss of species on land. Avian dinosaurs (chickens, crows, sparrows, hawks) escaped mass extinction in event number five (caused by the impact of an extraterrestrial object at the end of the Cretaceous). Ocean predators—ammonites, plesiosaurs, and mosasaurs—plus the non-avian dinosaurs disappeared. A while later, "Terror birds" stalked the predator-depleted landscapes of Paleogene time.[11]

10,000 Years from Now

Are these five, ancient episodes of turnover in the biota of earth a cautionary tale? Do they hold significance for science teaching? They seem to bring into question the stability of earth's life support systems. Elizabeth Kolbert, in voicing the research of many scientists (and winning a Pulitzer Prize), eloquently warns that the sixth mass extinction event in the past half billion years of earth history is upon us.

The rate of decline in biodiversity in the past several decades, extrapolated into the future and exacerbated by climate change, says so. The predicament of amphibians vanishing from habitats globally, but especially in the tropics, underscores this claim and stimulated Kolbert to write about the sixth (but first "unnatural") mass extinction.[12] In this one, the cause does not hide behind a veil of competing, conflicting, marginally supported hypotheses.

People are responsible in a multitude of ways: toxic chemicals, habitat destruction, resource exploitation, and greenhouse gas enrichment of the atmosphere. What does the precipitous decline in the forms of life on earth portend for the future? How might people respond? What form might this wicked problem take in a secondary school curriculum? Accounts of the first five really big extinction events provide insights in the complex interactions of evolution and environment. Forecasting scenarios on the other side of the sixth extinction are fraught with difficulty—but not impossibility. Fewer forms of life will inhabit the globe.

Hypothesizing other plausible effects offers fertile ground for student imagination grounded in reality. What signatures in the rock record might these effects leave behind? A team from the University of Copenhagen, Lene Møller Madsen, Robert Evans, and Rie Hjørnegaard Malm, partnered with Danish secondary schools to introduce students to thinking about just this sort of wicked problem—a first step toward assuming responsibility for stemming the tide of the sixth extinction.[13]

To make this problem tractable, the team reduced the challenge to predicting observations of strata geologists might uncover 10,000 years in the future above the ice age clays. What might these strata reveal about environmental change? How might fossils, chemistry, and sediment composition record change? Madsen, Evans, and Malm expected students to make predictions by analogy to the geology and its interpretation at Stevns Klint. "What remnants of today's world might become part of a boundary uncovered 10,000 years into the future?" they asked the students.

Many answers based on thoughtful analogies with past boundary depositions are possible. There are no correct predictions, only plausible and persuasive ones derived from knowledge of geological boundaries and climate change science. In response to the acceleration of habitat, climate, landscape, icescape, sea level, and extinction rates of change circa 2000, geologists have proposed defining a new epoch to the geologic timescale named the "Anthropocene," a successor to the previous summit of time, the Holocene.[14]

Hanna, Wyoming

The earth spiked a fever recorded at the Paleocene-Eocene boundary fifty-five million years ago in response to a greenhouse gas increase in the atmosphere.

So say the rocks of Hanna, Wyoming, interpreted by paleobotanist Ellen Curano of the University of Wyoming. Carefully dated strata captured a "thermal maximum event" that transformed ecosystems within 1,000 to no more than 10,000 years—geologic high speed that sent palm trees, turtles, and crocodiles scurrying north. On the far side of the boundary lived big warty beasts rather unlike most of the fauna of today.

This Paleocene-Eocene Thermal Maximum (PETM) is the best analogy in the geologic record to the extinction rates of the Anthropocene. For thousands of years the earth warmed at a rate of .025 degrees Celsius per century (5 degrees in all). The rates at present, however, are at least ten times as great, one to four degrees Celsius per century, with an upper limit unknown.[15]

Paleontologists debate the actual cause of the warming at the Paleocene-Eocene boundary—methane hydrates from the ocean floor and carbon dioxide from volcanic eruptions are among the competing hypotheses.[16] Regardless of which gas and manner of its release, the significance of the analogy remains unchanged. The PETM's gaseous burp roughly equaled in effect what burning all of the remaining carbon fuel reserves of today's earth would do.

The dawn of the Anthropocene must stand out in the geologic record to have meaning—a transition analogous to the Cretaceous-Paleogene boundary. What will be the analogue to enriched iridium signifying the change? Rapid extinction is a prime candidate. Strata anomalously enriched in plastics is another. According to Australia's national science agency, seafloor sediments now encase nine to fifteen million tons of microplastics (pieces from five millimeters to one-thousandths of a millimeter in size).[17]

Perhaps the onset of large-scale agriculture and livestock husbandry will do the trick—the bones of cattle, pigs, sheep, goats, and horses resting in a geologic formation sitting atop the last signals of elephants, rhinoceroses, giraffes, grizzlies, and musk oxen. The most precise marker under consideration is the burst in radioactivity that rained down in fallout starting with Hiroshima on Monday, August 6, 1945, at 8:16 a.m., enriched with the subsequent decades of nuclear weapons testing.[18] An extraterrestrial object can do no better.

Field study at Stevns Klint anchors the wicked problem of making a "geologic strata forecast." Madsen, Evans, and Malm posed a wicked problem for students to study stemming from the rock record in Denmark's cliffs. Other sites, such as Hanna, Wyoming, suggest parallel studies for connecting climate change to rapid extinction. In lieu of the expense and logistics of field study, classrooms may profitably use artifacts, photographs, and replicas of fossils to provide *direct experiences* for students. The *social problem* of climate change and the rising rate of species extinction is vital to solve.

A seaside cliff can be many things—from a pleasant childhood memory to the object of landscape art. Reconstructed educationally with new meaning, the experience of a seaside cliff becomes an acute awareness of the human capacity to affect change on a geologic scale. Disciplined understanding, put to work solving a wicked problem, inscribes the cliff with this meaning in an unmistakable manner.

A TIME OF WICKED PANDEMIC

The severe acute respiratory syndrome coronavirus 2 (SARS-CoV-2) pandemic has visited tragedy and suffering on millions of persons. By stretching health care systems to the limits of their capacities and squelching economic life, the Covid-19 crisis has rekindled outbreaks of other infectious diseases. Heroic citizens have risked and sacrificed their own lives to serve others; reprehensible charlatans have spread lies and conspiracy theories about its origin and cure. News focuses incessantly on medical data, prompting the public to examine its nature and, ultimately, recognize the challenge of clinical research with human subjects. Not since the launch of the Soviet communications satellite, Sputnik, in October 1957, and the American Apollo space program Moon landing in July 1969 has gazing at scientific data reached such an apogee.

Covid-19 made clear in 2020 that the citizens of the world were one humanity, under biological constraints, with inequality and injustice for many. The crisis has the potential to awaken the public to the need to educate students for lives of social responsibility, a fundamental aim for all subjects, science included.

Disease in the rearview mirror of history now bear down on present circumstances: 1918 Spanish flu, mortality on the march of Napoleon's Army to Russia and back again, the winter of suffering at Valley Forge, the Justinian Plague of the sixth century and the Black Death of the fourteenth (both transmitted by bacterium, *Yersinia pestis*), and the depopulation by bubonic plague of the Atlantic seaboard communities a few years prior to the arrival of the Pilgrims, just one episode in the pandemic-driven demise of indigenous populations throughout the Americas.[19]

There is no understanding of what is happening to society or what might be the long-term solution without insights from cell and molecular biology, immunology and epidemiology, sociology and social psychology, climate change and ecology. The role of technology in response to a twenty-first-century pandemic merits critical examination. Neither has the mathematics of exponential growth, conditional probabilities, and sampling distributions been more relevant to public life, nor have disparities between advantaged and disadvantaged populations more starkly played out.

Balcony singers in Italy inspire everyone to remember that song fights gloom even when Zoomed; the arts, as much as the sciences and politics, are needed to rise to meet the challenges of a global pandemic, this time and the next. Maybe, just maybe, out of the chaos, fear, confusion, and suffering circling the globe, educators may fashion new, enduring, growing, and evolving institutions of learning—ones that develop curricula that are timely, local, and particular and responsive to social need.

History offers several precedents: the integration of art, writing, and topic work in British primary schools following the Second World War, Maria Montessori's adaptations of early childhood education to serve working-class Italians at the start of the twentieth century, and Anna Botsford Comstock's classic *Handbook of Nature Study* intended to rejuvenate life in rural New York in the wake of the Panic of 1893.

Now is the time for all good science teachers to question traditional achievement and engage youth in learning to care for the community. Lessons may begin with how the chemistry of soap disarms SARS-CoV-2.[20]

NOTES

1. Horst W. J. Rittel and Melvin M. Weber, "Dilemmas in a General Theory of Planning," *Policy Sciences* 4, no. 2 (1973): 155–169. Wicked problems have no immediate and no ultimate test of a solution; they are never solved but constantly re-solved. Climate change is such a wicked problem and needs to be addressed through education that acknowledges this complexity.

2. Marianne Achiam, Melissa Glackin, and Justin Dillon, "Wicked Problems and Out-of-School Science Education: Implications for Practice and Research," in

Addressing Wicked Problems: The Role of Out-of-School Science Education, eds. Justin Dillon, Marianne Achiam, and Melissa Glackin (Berlin, Germany: Springer Nature, 2021), 229-237, doi: 10.1007/978-3-030-74266-9.

3. Andrew Blum, *The Weather Machine: A Journey Inside the Forecast* (New York: HarperCollins Publishers, 2018), 70.

4. Ibid., 126.

5. Naomi Oreskes, *Why Trust Science?* (Princeton, NJ: Princeton University Press, 2019).

6. Elizabeth Kolbert, "What Will Another Decade of Climate Crisis Bring?" *The New Yorker* (January 13, 2020), accessed August 17, 2020, https://www.newyorker.com/magazine/2020/01/13/what-will-another-decade-of-climate-crisis-bring.

7. Intergovernmental Science-Policy Platform on Biodiversity and Ecosystem Services (IPBES), "Nature's Dangerous Decline 'Unprecedented'; Species Extinction Rates 'Accelerating,'" United Nations Report, Sustainable Development Goals, accessed August 17, 2020, https://www.un.org/sustainabledevelopment/blog/2019/05/nature-decline-unprecedented-report/.

8. James Lawrence Powell, *Night Comes to the Cretaceous: Dinosaur Extinction and the Transformation of Modern Geology* (New York: W. H. Freeman and Co., 1998).

9. Tuve Damholt and Finn Surlyk, "Stevns Klint—Ny Dansk Verdensarv," *Geoviden* 3, 2–19, cited in Lene Møller Madsen, Robert Evans, and Rie Hjørnegaard Malm, "Using a Wicked Problem for Inquiry-Based Fieldwork in High-School Geology: Addressing Climate Change and Mass Extinction Events," in *Addressing Wicked Problems*, eds. Justin Dillon, Marianne Achiam, and Melissa Glackin (Berlin, Germany: Springer Nature, 2021), 167-188, doi: 10.1007/978-3-030-74266-9.

10. David M. Raup and Steven M. Stanley, *Principles of Paleontology*, 2nd ed. (San Francisco: W. H. Freeman and Co., 1978), 439, 448.

11. John Alroy, "Colloquium Paper: Dynamics of Origination and Extinction in the Marine Fossil Record," *Proceedings of the National Academy of Sciences (PNAS)* 105, no. suppl 1 (2008): 11536–11542, https://doi.org/10.1073/pnas.0802597105. David M. Raup and John J Sepkoski Jr., "Periodicity of Extinctions in the Geologic Past," *PNAS* 81, no. 3 (February 1984): 801–805, https://doi.org/10.1073/pnas.81.3.801.

12. David B. Wake and Vance T. Vredenburg. "Are We in the Midst of the Sixth Mass Extinction? A View from the World of Amphibians," *PNAS* 105, suppl 1 (August 12, 2008): 11466–11473, https://doi.org/10.1073/pnas.0801921105.

13. Madsen, Evans, and Malm, "Using a Wicked Problem for Inquiry-Based Fieldwork in High-School Geology: Addressing Climate Change and Mass Extinction Events."

14. Richard Monastersky, "First Atomic Blast Proposed as Start of Anthropocence," *Nature News* (January 16, 2015), accessed August 17, 2020, https://www.nature.com/news/first-atomic-blast-proposed-as-start-of-anthropocene-1.16739.

15. Lee R. Kump, "The Last Great Global Warming," *Scientific American* 305, no. 1 (July 2011): 61.

16. Emily Graslie, "Tiny Teeth, Fearsome Beasts," *Prehistoric Road Trip*, episode 3 (Public Broadcasting System [PBS], July 1, 2020), https://www.pbs.org/video/tiny-teeth-fearsome-beasts-axy0yx/.

17. Justine Barrett, Zanna Chase, Jing Zhang, Mark M. Banaszak Holl, Kathryn Willis, Alan Williams, Britta D. Hardesty, and Chris Wilcox, "Microplastic Pollution in Deep-Sea Sediments from the Great Australian Bight," *Frontiers in Marine Science* 7 (October 5, 2020), accessed October 14, 2020, https://doi.org/10.3389/fmars.2020.576170.

18. The Trinity Test of an atomic bomb occurred on July 15, 1945. From a geologic perspective, decades of testing would create a stratum enriched with isotopes from radioactive fallout.

19. Nathaniel Philbrick, *Mayflower: A Story of Courage, Community, and War* (New York: Viking Penguin, 2006), 48.

20. Jonathan Corum and Carl Zimmer, "How Corona Virus Hijacks Your Cells," March 13, 2020, https://www.nytimes.com/interactive/2020/03/11/science/how-coronavirus-hijacks-your-cells.html?searchResultPosition=2.

Chapter 27

Harvesting Oysters

Look upon the landscape less as its "possessor" and more as its "companion" in a meaningful journey, teaches the nature writer Barry Lopez. Achieving this perspective requires committing to long-term, multigenerational residency, the cultivation of local knowledge, and the "opening of an intelligent conversation" with the land.[1]

Knowing who one is and where one stands, literally and metaphorically, bestows strength. An intimate knowledge of place clarifies "what is inviolate about the relationship between a people and the place they occupy."[2] Bodies and minds depend upon particular places for sustenance. Memories of important events, whether signified by a monument in the city or a landform in the wilderness, attach individuals to each other, their social histories, and their collective identities.[3]

"Place-based" education thrives as an antidote to "standards-based" reforms. Achieving a sense of place walks in tandem with finding membership in a community. Together, they are the principal ingredients of a resilient self.

David Gruenewald has admonished educators for according low priority to achieving a sense of place: "Fundamentally significant knowledge is knowledge of the unique places that our lives inhabit; failure to know those places is to remain in a disturbing sort of ignorance."[4] A pedagogy of place values becoming native to a place. Organizing instruction in harmony with achieving a sense of place melts away distinctions among subjects, cultivates identity as much as knowledge, and fosters acting to improve the community.

INDIGENOUS KNOWLEDGE

In many respects, indigenous peoples are the consummate sense-of-place experts. In time, any people may come to feel native to their place of being. Stories told by third- and fourth-generation descendants of Europeans, for example, often converge on themes expressed by indigenous authors. According to historian Elliot West, as the children of pioneers become native to a place, themes of new beginnings and starting over give way to gratitude for sustenance and to an appreciation of the power of the landscape to cultivate spiritual well-being.[5]

Gregory Cajete (Tewa), in his seminal 1994 book, on indigenous educational traditions, *Look to the Mountain: An Ecology of Indigenous Education*, summarizes how:

> From methods of navigation, to application of medicinal properties of plants and animals, to traditional techniques of agriculture, . . . Indigenous people have demonstrated a way of knowing and relating that must be regained and adapted to a contemporary setting—not only for the benefit of those cultures themselves, but for all humankind.[6]

Tribal members often have an acute sense of the need to preserve historic and sacred places as well as resources. Well-managed Indian lands and resources depend upon Native American professionals with perspectives drawn from both tribal knowledge and scientific expertise.[7] Tlingits, the same as in many tribal communities, learn to understand the behavior and roles of creatures in an ecosystem and use this knowledge to guide their interactions with the environment. Tlingit youth gain this knowledge not in a classroom but rather from elders through oral histories, songs, crafts, and practical training.[8]

Tsé na'alkaah

Geologist Steven Semken perceives an overemphasis on global syntheses in mainstream geoscience education that displaces concerted study of local places. For indigenous students, such places often have significant prior meaning that must be respected in their cultures. In some instances, geoscience depicts culturally significant places inappropriately.

In Diné, the language of the Navajo nation, "tsé na'alkaah" means "rock study." The term is also the translation of a course title, "Indigenous Physical

Geology," taught by Semken to Navajo Nation students on their land. The course translates place names and geologic terms into Diné and includes topics such as climate, environmental quality, geology of the Colorado Plateau, and tectonic activity along the western boundary of the North American plate. Students explore factors that sculpt regional topography, study the drainage pattern of the Colorado River, and learn about the origins of uranium deposits and coal beds on the plateau.[9]

In conventional geoscience teaching, cancer rates and liver disease fall outside the boundaries of course syllabi. Among Navajo uranium miners, cancer rates measured in the 1980s rose to more than fifty times the national average and liver cancer rates even higher.[10] Geology, economy, health, and environment are inseparable concerns in their education, and attachment to place fosters their understanding.

To counter omission and misrepresentation and act upon the meaning of place, Tsé na'alkaah students make presentations to local audiences about groundwater contamination from uranium mining, flyash disposal at coal-fired plants, and problems associated with decadal scale drought.[11] Their knowledge overcomes the detriment of attention to local circumstances given by global syntheses of geology.

Salmon Camp

Women in Native American communities have long held the traditional role of "keepers" of scientific and traditional ecological knowledge essential for subsistence living. This role contrasts with low employment rates of Native American women in science, technology, engineering, and mathematics. There is clearly an unrealized potential for Native American women to become leaders in the fields of science, mathematics, and engineering—pathways requiring a very significant decision.[12]

How might this pathway become more inviting and accessible to Native American youth? Dual guidance by practicing scientists and tribal elders holds promise. Field investigations, for example, readily link concerns felt by tribes in the Pacific Northwest for the sustainability of fish harvests and fish habitats to ongoing research. Collaborating networks of students, families, elders, and scientists places science in a social context significant to Native Americans and others intent upon restoring places valued for their role in sustaining a community:

The challenge is daunting, but not by any means insurmountable—at least when Native American youth, equally led by tribal members and elders, and scientists from both within and beyond indigenous communities, immerse themselves in out-of-school research experiences integrated with the study of tribal knowledge. The sciences can, in the company of wisdom born of multigenerational residency, look through time to trace out the ligaments binding landscapes, past and future, to culture and identity.

Local knowledge is essential to understanding local ecological problems. Local tribal knowledge specific to tribal homelands "is relevant to the daily struggle of a particular people."[13] Local knowledge understood deeply means to become "intimate with the land,"[14] the starting point for developing a sense of place and a primary objective of "Salmon Camp," a long-standing program for Native American youth currently under the auspices of the Columbia River Intertribal Fish Commission. The commission represents the interests of the Yakama Nation, Confederated Tribes of the Umatilla, Confederated Tribes of the Warm Springs, and Nez Perce Tribe.[15]

Native youth at Salmon Camp grow to appreciate place in four senses. First, they recognize place (a) as "homeland" and second, after field study, as (b) specific places to return to cultivate intimacy with the land. Their sense of place encompasses (c) the Pacific Northwest as a wondrously distinct region shared by many peoples. The Salmon Camp community resides in their hearts as (d) a memory to revisit as a source of resilience and perseverance in support of career and community service aspirations. Salmon Camp creates a safe, empowering place to be "Native and crazy and tearful and a scientist and a fish geek"—all at once.[16]

Bearing witness to place as homeland, understanding the meaning of salmon as a cultural heritage, and learning the technological skills needed to conserve and restore their abundance requires the blending of indigenous and nonindigenous ways of knowing nature. Looking through this lens changes perspective. Vehicles speed along the interstate highway through the Columbia Gorge, passing the small village of Celilo. For Salmon Campers, the village symbolizes endurance, tradition, commitment. Nearby Celilo Falls was a traditional fishing and gathering place for several thousands of years. Indigenous fishers still sell their catches at roadside booths and the back of pickup trucks parked in the town of Cascade Locks on the shore of the Columbia River.

The consummatory experience of meaningful attachment to landscape joined to the satisfying feeling of membership in a community strengthens identity, whether of an indigenous or immigrant people. With confident identity comes commitment to community betterment.

Most any place will do, interpreted with the proper stories.

SKAMOKAWA

Far downstream from Celilo, the dilapidated town of Skamokawa, Washington, faces the broad Columbia. The name means "smoke on the water"—fog shrouding the landscape. Here vibrant communities have come and gone, shifting from one resource to another and adapting new technologies. Once Skamokawa was a Venice-like community made prosperous by the cannery, dairy, and timber industries. Rowboats were the favored means of commuting to school.

Other towns of the Lower Columbia have vanished, both Chinook and European. At one Cathlamet village site, cedar lodges housed several hundred people. Occasionally, blue trade beads from 200 years ago turn up on the shores near Skamokawa. "Until I learned these stories, I thought Skamokawa was just another pleasant place to bike through and maybe stop for a sandwich," remarked a teacher attending a weekend seminar on achieving a sense of place offered through Lewis & Clark College.

The songs of "A Portrait of Place" performed by The Willapa Hills band celebrate the peoples and lands of the Lower Columbia. They play waltzes honoring Swedish and Finnish immigrants. They sing of conflict in the 1880s over salmon seining grounds. A haunting set of lyrics tells of women harvesting oysters during Second World War blackouts—women who counted on their neighbors' horn-honking to guide them back to shore from the oyster beds in tidal flats.[17]

Students from Wahkiakum County, Washington, have surveyed the banks of the Columbia for invasive, non-native species, searched beds of moss for crane-fly larvae, collected seafloor fossils from hillsides lifted by earthquakes, and observed endangered Caspian terns feasting on endangered runs of salmon smolts—a source of environmental policy dilemma. Strata of geological, historical, and ecological stories formed the curriculum. Analogous lessons—leading to escape from a "disturbing sort of ignorance"—can be learned anywhere when teaching timely, local, and particular sciences.

Good things happen when bettering the community supersedes alignment with standards, and education, respectful of how cultures embed themselves in places, goes beyond standards. Disuniting the sciences, then selecting knowledge deemed worthwhile for achieving resiliency, sustainability, and prosperity, can play an important role for indigenous peoples recovering from the era of colonial dispossession. At the same time and in much the same way, achieving a sense of place and responsibility for future generations' well-being can inculcate humility among privileged peoples.

NOTES

1. Barry Lopez, *The Rediscovery of North America* (New York: Random House, 1992), 40.
2. Ibid., 36–37.
3. Edward S. Casey, "Public Memory in Place and Time," in *Framing Public Memory*, ed. Kendall R. Phillips (Tuscaloosa, AL: University of Alabama Press, 2004), 17–44.
4. David A. Gruenewald, "Place-Based education: Grounding Culturally Responsive Teaching in Geographical Diversity," in *Place-Based Education in the Global Age: Local Diversity*, eds. David A. Gruenewald and Gregory A. Smith (Mahwah, NJ: Lawrence Erlbaum, 2008), 143.
5. Elliot West, *The Way to the West: Essays on the Central Plains* (Albuquerque, NM: University of New Mexico Press, 1995).
6. Gregory Cajete, *Look to the Mountain: An Ecology of Indigenous Education* (Durango, CO: Kivaki Press, 1994), 79.
7. Janine Pease-Pretty On Top, "The Last Word: At the Heart of Indian Country," *Winds of Change* 20, no. 1 (2005): 78.
8. Alaska Native Knowledge Network, "Herman Kitka: Traditional Ecological Knowledge," accessed August 17, 2020, http://ankn.uaf.edu/Resources/course/view.php?id=11.
9. Steven Semken, "Sense of Place and Place-Based Introductory Geoscience Teaching for American Indian and Alaska Native Undergraduates," *Journal of Geoscience Education* 53, no. 2 (March 2005): 149–157.
10. Oscar Schwartz, "A Geneticist's Dilemma," *Washington Post Magazine* (November 23, 2020), accessed November 30, 2020, https://www.washingtonpost.com/lifestyle/magazine/.
11. Semken, "Sense of Place," 154.

12. Pease-Pretty On Top, "The Last Work," 78.

13. Gregory Cajete, quoted in "From 'Savages' to Scientists: Mainstream Science Takes First Step toward Recognizing Traditional Knowledge," *Tribal College Journal* 15, no., 1 (2003): 12.

14. Lopez, *Rediscovery of North America*, 34.

15. Columbia River Inter-Tribal Fish Commission, accessed August 17, 2020, https://www.critfc.org/.

16. Phyllis Campbell Ault, "Salmon Cycles: Influences of a Science Field Study Immersion Experience with Native American Young Women," EdD diss., (Lewis & Clark Graduate School of Professional Studies, 2007), iv.

17. Ruby Murray, "Willapa Hills' Portrait of Place, A Labor of Love," *Wahkiakum County Eagle* (February 3, 2011), accessed August 17, 2020, https://www.waheagle.com/story/2011/02/03/news/willapa-hills-portrait-of-place-a-labor-of-love/1737.html.

Chapter 28

Caring Communities

The children at Arbor School of Arts and Sciences engineered automata from simple machines to enact stories representative of a year of study. Middle school students in Seneca County, New York, launched weather balloons laden with air quality sensors to the stratosphere. The data collected contributed to crowdsourced climate modeling. At Arbor, celebration and cooperation strengthened a community of learners. In New York, collaboration with community partners gave authenticity and stature to student work.

Concern for cognitive achievement typifies conventional schooling. In a conventional classroom, children learn types of simple machines and the concept of mechanical advantage. In a traditional secondary (or middle) school earth science course, students assimilate knowledge of atmospheric physics and chemistry useful to weather forecasting. What is there to learn at a more fundamental and lasting level exemplified by Arbor's automata and Seneca County's ballooning?

From a Deweyan perspective, the answer is clear: the *value of the experience itself*. Arbor children learned the *value* of synthesizing and celebrating the theme of invention and discovery. New York students learned the *value* of collecting real weather data useful to the local farming community. For a learner, "value" is the feeling of significance, the bridge between volition and cognition, a value dependent upon social interaction. Disassociation of art, literature, science, and history from social contexts is miseducative. Ideally, wrote John Dewey, a school is a "miniature community . . . in close interaction with . . . experience beyond school walls."[1]

CAMS TO MAPS

The Arbor Intermediates (grades four and five), adhered to a theme of study, engineered captivating automata, and celebrated their accomplishments. They collaborated in small groups across grade levels in pursuit of the history of discovery and invention. Their schooldays respected the integrity of traditional subjects while integrating their pursuit with project work. Mechanical arts mattered as much as literary ones. They applied mathematics and science to the challenge of creating automated actions representative of significant learning during the course of the year. The technology of cams and levers symbolized an understanding of simple machines.

The science they learned proved its worth. Most importantly, the curriculum fostered the creation of a purposeful community at the classroom level, the teachers ever mindful of creating community from the interactions among students and with subjects.

Partners in Farming

Fortune favored the rural schools of central New York. Partners from the community reached out to engage adolescents in the work of science promising benefits to their farming families. High-altitude ballooning arriving by Weather Truck mesmerized middle schoolers while at the secondary level modeling inspired by crowdsourcing principles put the work of amateurs in competition with professional forecasting. Air quality eggs came to school to embed meteorology (and climate science) in community purpose.

The Bootleg Balloon program nested schools, families, residents, a nature center, a museum, and a university in a series of caring communities. Meteorological investigations bound students together in learning science helpful for solving the farming community's timely, local, and particular problems. Interacting with online databases connected participants to a wider digital community. Developing an intuitive sense of atmospheric phenomena augmented understanding of how climate scientists make sense of complex systems. This knowledge, predicated upon the close correspondence between meteorological inquiry and climate science, is essential to establishing trust in scientific work of value to the regional community.

RESPONSIBILITY AND EQUITY

Programs such as Bootleg Ballooning promote educational opportunity among disadvantaged populations in communities underserved by cutting-edge science teaching. Where social history has hindered opportunity, prioritization of resources to promote equity is a moral and political imperative; for example, in economically stressed rural areas and agricultural communities facing the impacts of climate change.

Competitive Equity

Injustice, unfairness, discrimination, and inequality accumulate to degrade community life. Creating opportunity and intervening to equalize opportunity regardless of race, gender, class, or ethnicity is the key to individual advancement and community improvement. No one of good will and clear reason questions the value of equity. Care must be taken, however, not to reduce this value to a single-minded notion of equalizing competition among individuals, whether for jobs, education, wealth, or power. "Competitive equity" indeed deserves support, but not without equal devotion to "reciprocal equity."

Schooling emphasizes measurable achievement and standardizing makes levels of achievement legible, a prerequisite of measurement.[2] Unfortunately, the regimentation that necessarily follows, and the recourse to competition as a motivator, easily obscures the basic functions of education in a democratic society. Public education's primary purpose is to develop citizens dedicated and able to live lives of social responsibility. Democratic society encourages citizens to exert control of the circumstances governing their lives with understanding, compassion, and reason.

Students do not attend schools while waiting to transmogrify into citizens once they have acquired a competitive set of skills. Schools are communities composed of young citizens learning through experiences the meaning of responsibility to self, each other, and the world outside the classroom. Children and adolescents are citizens being nourished to think, feel, and act.

Reciprocal Equity

Zeal in the measurable pursuit of lifting achievement, expressed ethically as equalizing educational opportunity while promoting excellence, reinforces competition among individuals. Competitive equity, opportunity, and excellence coexist and reinforce each other. Their satisfactory accomplishment is, to state the obvious, vexing. Inequity and inequality stubbornly resist eradication; the "achievement gap," for many reasons, persists.[3] The "market paradigm" casting citizens as producers and consumers, competing for the fruits of prosperity, offers fertile ground for devotion to the value of competitive equity.

Accepting as a fundamental educational aim the cultivation of socially responsible living in a democratically governed society calls for reconceptualizing equity. The shift begins with a shared commitment to mutuality. Society and individuals have a mutual interest in maximizing individual potential and removing unfair obstacles to its attainment. At the same time, individuals and society are locked into a reciprocal relationship. Reciprocity is the central component of a human community. Reciprocity, in turn, requires trust. People do things for each other when they trust each other to reciprocate, often with gratitude.

"Reciprocal equity," as an expression of this mutuality and "caring" rather than "competing," characterizes its nature. Everyone profits from the cultural heritages and natural resources not of their own making that sustain and enrich their lives. In the interests of the future of the community, each person incurs the responsibility to reciprocate for the unearned forms of grace bestowed upon them. "Gratitude" permeates the unfolding of reciprocal equity—gratitude by the individual for the gifts from the community and gratitude by the community for the talent shared by each individual.

According to Robin Wall Kimmerer (Potawatomi), the sense of equity embedded in this reciprocity matches educational aims associated with indigenous ways of knowing "rooted in intimacy with a local landscape where the land itself is the teacher."[4] From intimacy with the land, inhabitants learn to respect how each being, according to its nature, exists in a giving and reciprocal relationship with others. The web of these relationships gives strength to the community for which each member feels gratitude. Reciprocal equity enacts a sense of place.

These insights, believes Kimmerer, teach children to recognize their own gifts, to realize how their talents might be of service to the community, and to feel gratitude for the gifts received from others, just as the forest gives thanks for the moss and the moss gives thanks for the rain. "Whatever our gift," Kimmerer writes, "we are called to give it and to dance for the renewal of the world in return for the privilege of breath."[5]

The sense of reciprocity found in contributing to the life of a community, vital to responsible living in a democratic society, enhances the ideal of equity. Neither sense of equity—competitive or reciprocal—in the absence of the other holds sufficient social promise. Both senses bear upon science teaching.

In terms of reciprocal equity, the individual is to the community as the school is to society. A school may claim status as "progressive" by prioritizing the development of individual talents, yet, unless adhering to a curriculum disciplined by social goals, falls short of the ideal of reciprocity.[6] "Disciplined" in this context means constructing knowledge to achieve social value, such as improving local forecasting in the farming communities of central New York. Subject matter, matters—and, from one time and place to another, the subjects that matter may differ. No career better exemplifies their enactment than that of Neal Maine.

Citizen Maine

Neal Maine taught both fourth-grade and high school biology to students from Oregon's northern coastal communities. Raised in Clatsop County, he never left. Instead, he brought attention to his community and taught generations of students to respect its beauty, appreciate its capacity for renewal, and value its gifts to the well-being of the community. Maine became his district's science supervisor and the state-appointed coordinator for a natural resources curriculum. In the latter years of his career, he received an Annenberg Foundation Rural Challenge grant and became a steward of the North Coast Land Conservancy.

Across from Seaside High School stands testimony to his legacy: a viewing deck and estuary access. From this access point, thousands of students have conducted inquiries about the mouth of the Necanicum River, the Seaside bar, and the local intertidal zone. They have censused ghost shrimp, mapped barnacle encrustations, and measured the range of salt tolerant sedges. They learned science in partnership with the community and collected data respected by professional ecologists.

One high school group applied their trigonometry skills to the hazards of tsunamis to determine areas of risk based on elevation. They presented proposed pathways for evacuation to the Seaside City Council.

In another project, students investigated the buying habits of their peers who often went to Portland to shop, then shared this information with local merchants as part of a "buy local" campaign. Other classrooms have studied how to make playgrounds more attractive to young children and to promote the display of school art in public places.

Maine's cadre of citizen scientists keeping records of dead common murres washing ashore on local beaches, several years ago, noted a spike in adult mortality. The spike signaled something was amiss at sea. The "something amiss" turned out to be an oil spill that even the Coast Guard had, for a time, missed.

Maine always reminds anyone who will listen that schooling is not a drawn-out apprenticeship with the reward of citizenship at its end. Adults may have rights that children lack (such as voting), but being ten or fifteen years old does not disqualify a person from citizenship and community membership. The point of curriculum, argues Maine, is to gain access to resources for interacting and participating in the world, to empower citizens—children and adolescents included.

He dreams of creating "Young Citizen Development Centers" devoted to community building, either within or independent of schools. Maine's concept of citizen participation—and of the aim of educating people for productive lives of social responsibility in a democratically governed society—invigorates education, exemplifies the meaning of reciprocal equity, and fosters caring communities.[7]

NOTES

1. John Dewey, *Democracy and Education: An Introduction to the Philosophy of Education* (New York: MacMillan, 1922), 418.

2. "Legibility" is a central concept in James C. Scott, *Seeing Like a State* (New Haven, CT: Yale Press, 1998). Scott develops this notion with a parable about a forest. In the parable foresters replace a wild forest of many species with a single species forest planted in rows for efficient measurement and management. In *Challenging Science Standards* (Lanham, MD: Rowman & Littlefield, 2015), I retell this parable in terms of standardizing curriculum. "Legible" forms of achievement enable the comparison of learning outcomes.

3. Obed Norman, Charles R. Ault, Jr., Bonnie Bentz, and Lloyd Meskimen, "The Black-White 'Achievement Gap' as a Perennial Challenge of Urban Science Education: A Socio-Cultural and Historical Overview with Implications for Research and Practice," *Journal of Research in Science Teaching* 38, no.10 (December 2001): 1101–1114.

4. Robin Wall Kimmerer, "*Gathering Moss: A Natural and Cultural History of Mosses* (Corvallis, OR: Oregon State University Press, 2003), 101.

5. Robin Wall Kimmerer, "The Giveaway," in *Moral Ground: Ethical Action for a Planet in Peril*, eds. Kathleen Moore and Michael Nelson (San Antonio, TX: Trinity University Press, 2010).

6. David Hawkins, "John Dewey Revisited," in *The Informed Vision: Essays on Learning and Human Nature* (New York: Algora Publishing, 2002), 165.

7. Charles R. Ault Jr., "Achieving Querencia: Integrating a Sense of Place with Disciplined Thinking," *Curriculum Inquiry* 38, no. 5 (December 2008): 605–637.

Conclusion

Paradigm's Progress

Use of the adjective "scientific" raises little concern in everyday conversation; ditto for "unscientific." Their easy acceptance, however, masks disagreements over what to teach widely and why. The sciences marshaled to understand the Cascadia subduction zone differ from those assembled to examine problems of seawater incursion in low-lying communities in Florida, the decline of soil moisture in the upper Midwest, and the toxicity of soil on Native American lands.

Perhaps "science is as science does"—and that is what we ought to teach. Or maybe it's not that simple.

Even if science is as science does, do all the sciences do science equally scientifically? "My science is as good as your science" each discipline claims, advancing a self-serving argument of deserving a seat at the curriculum table. When the Pandora's box of disunited sciences opens, along with its companion chest of how the sciences matter to economic competitiveness, national security, and social opportunity, do science educators, the same as Forest Gump opening a box of chocolates, not truly know what they're gonna get?

STEREOTYPICAL SCIENCE

The "oppressive rhetoric modeled on physics," inherited from the Cartesian paradigm, has dominated the sciences[1] and echoed through the corridors of school science for a very long time. In keeping with Descartes' agenda, "most physicists dream, above all of a uniquely self-consistent mathematical theory that explains and determines all the microphysical constants, as well as the entire cosmic evolution."[2] A "theory of everything" would account for the charge on an electron and the expansion of the universe while reconciling quantum mechanics with relativity theory. There's nothing wrong with this aim, except as a standard for explaining most phenomena of interest to humanity.

National standards preach adherence to three dimensions of science for all Americans: practices, concepts, ideas. These are words neutral in value that embody a lost paradigm of science, yet they dominate the current instructional paradigm of science teaching. School policy makers expect science educators to align each lesson and every assessment with the dimensions of the standards. The alignment comes at a modest price: separating matrices of content from methods of inquiry responsive to the nature of the case.

Alignment pays an even higher price: dismissing the timely, local, and particular circumstances of history as guidance for curriculum design responsive to social values.

The stories of science beyond standards make arguments that threaten to render achievement illegible to traditional measures. That is exactly what happens when a promising paradigm displaces a senescent one. Established criteria of evaluation, once widely depended upon, no longer apply. Rubrics adapt or perish. Incommensurability happens; education evolves.

The Recurring Obsession

Simon Schama characterized the landscapes of our planet as "a text on which generations write their recurring obsessions. . . . All our landscapes, from the city park to the mountain hike, are imprinted with our tenacious, inescapable obsessions."[3] Schama sought to uncover the origins of obsessions for interpreting nature that have come to guide ecological restoration.

A simple substitution of "science" for "landscape" underscores the aptness of his insight to examining the perpetual desire to revitalize and unify science teaching. Among science educators, conceptions of teaching science as method, process, nature, or practice has been a recurring obsession, a cornerstone of successive paradigms. The time has come for science educators to look beyond their own professional habits of mind in order to escape the perpetual unification machine.

Every generation seems obsessed with its era's unprecedented threat and mobilizing the nation in response repeatedly turns to reforming science teaching. Fear of Russians and their launch of Sputnik goaded American science education into action in the 1950s and 1960s, as had world war in the 1940s. In the 1980s, the threat came from Japan and was economic. In the 2000s, it came from China—economic at first, then technological, and increasingly strategic.

The Great Pivot

The Second World War transformed the position of federal funding for science and the importance of science in the national consciousness. Achieving strategic and technological advantages during the Cold War made effective schooling in science and mathematics of utmost importance. Scientists strove to convince politicians, educators, and the public that methods of producing and generating knowledge adapted discipline by discipline to particular contexts in specific ways. In the guise of simple steps, scientific methods extrapolated to everyday life diluted their character.

From historian John Rudolph's perspective, science education pivoted dramatically when intellectuals began to wage war on teaching "the" scientific method.[4] Those who prosecuted this war won all of the scholarly battles over national policy. Victory, however, in the public mind and classroom consciousness, eluded them. The obsession to instill in every generation an understanding of the scientific method refuses to retreat.

The Fool's Errand

Separating method from content is a fool's errand. Content itself is a tool of inquiry. Concepts are engineered and tested for their worth; methods of inquiry are symbolic of understanding. Processes of scientific inquiry both produce and justify knowledge. In becoming continuous, inquiry recycles understanding.

A proper reading of John Dewey respects his understanding of disciplines as resources for solving societal problems. This perspective checks the tendency to conceive of designing experiments, collecting and interpreting data, and reflecting on the significance of findings as the principal content of scientific literacy. The truth—responsive to the nature of the case—is more complicated.

Descartes' dream is no longer adequate to the demands of present circumstances, and those who rightly characterize science-as-inquiry must take care not to perpetuate the discredited stereotype of scientific unity. Humans, scientists included, are not gods. With reason, they constrain uncertainty and reduce ambiguity about how the natural world works. When acting in concert, the sciences tolerate not only plurality but embrace diversity. Omniscience, like certainty, remains beyond reach. Often being "intelligently wrong"—a useful stage of progress and a characteristic of children's thinking—serves as a platform for seeing what needs fixing.[5]

The sciences simply cannot offer "freestanding and universal" methods of generating knowledge. Fortunately, rationality lives on, restricting its truths to context.

CULTIVATING TRUST

The stereotyping of science as abstract method conducive to finding objective truth seduces school science and the public. As a result, people doubt the credibility of any science that does not seem to "play by the rules," and sciences wedded to a historical style or dependent upon mathematical modeling encounter exaggerated skepticism. Burdening evolution and climatology with suspect scientific status invites religious dogma or political ideology to dismiss their findings. Cynicism and apathy grow within the voids created where unjustified suspicion trumps well-warranted claims.

In the midst of political debate, things that matter culturally compete with scientifically verified claims. Science teaching needs to garner public trust and encourage access to reliable information needed for making political judgments. Connecting science to everyday needs, hardly a new idea, seems a necessary starting point for earning the public's respect for scientific enterprises. Stereotyped science, however mesmerizing at first, ultimately makes the task difficult.

Backlash

If there is a silver lining in the dark cloud of the suffering wrought by the global Covid-19 pandemic, it may well be a call to improve public understanding of scientific diversity, and, thus come to mark, as did Sputnik, a new era in the history of science for all Americans. Humanity is desperate for a scientifically trustworthy vaccine, yet no doubt anti-vaxxers stand ready to propagate conspiracy theories intended to defeat well-warranted programs of immunization. Covid-19 has placed trusting the sciences on trial, and the well-being of billions of humans depends on the outcome.

Will peoples around the globe appreciate and accede to authoritative expertise in light of the pandemic, or will they react as did many in the United States following the horrors unleashed by the First World War and the subsequent Spanish flu pandemic? In *These Truths*, historian Jill Lepore recounts how fundamentalist Christian sects accelerated their growth postwar by blaming Godless secularism—the lamented outgrowth of evolution's march through schooling and society—for the chaos, evil, horror, and suffering in Europe.

Appeals to "higher" truths prompts people to discount scientific claims at odds with cherished values. Religious fundamentalists proselytized truths believed superordinate to the supposed facts of science. They held these truths to be self-evident, truths, in their minds, with implications for how best to order society, to align the "polis" with the "cosmos," to guide earthly life with heavenly ideals.[6]

Hope

William Jennings Bryan—of Scopes trial infamy to science educators, three times the democratic candidate for president and President Wilson's first secretary of state—championed both fundamentalist religion's agenda and Christian compassion's underpinning of labor law reform. Critical and fearful of eugenics (near-consensus science disgraced and repudiated in the aftermath of Nazi genocide) and opposed to discrimination targeting immigrants from southern and eastern Europe, Bryan's Democratic Party politics stood on the progressive side of social history even as he led the political crusade against Darwinism.[7] Bryan held faith in a curious hybridization of social justice and anti-evolutionism. Beliefs, not just truths, are complicated.

The pandemic dread of 2020 has turned with hope to the skills of computer modelers and vaccine engineers. Public safety depends upon algorithms and molecular biology. State, local, and government policy makers respond daily to the coders' forecasts; politicians promise better living through vaccinations.

In light of the epidemics of conspiracy theories and misinformation pervading twenty-first-century life inflamed by social media accelerants, earning trust is of the essence. Naomi Oreskes warns that the stakes are high, admonishing, "If we fail to act on our scientific knowledge and it turns out to be right, people will suffer, and the world will be diminished."[8]

Most people respect the threatening reports of climate change. Many worry about climbing extinction rates. The decline in marine and terrestrial biodiversity threatens public health, agricultural productivity, and spiritual enrichment. The public should not demand facts found only by controlled experimentation in order to trust expert advice on how to address these problems. The cultivation of trust calls for deeper wisdom, for an embrace of scientific plurality.

And something more: a decent respect for the Native American notion of "reciprocal equity" and the virtue of gratitude. The sense of reciprocity taught by Gregory Cajete (Tewa) and Robin Wall Kimmerer (Potawatomi) means committing one's talents to the good of creation in return for the gift of breathing—and many others.

STORIED MEANINGS

The students appearing in *Beyond Science Standards*, whether suburban fourth graders or Chinese graduate students, engaged in playful exploration, found aesthetic pleasure, strove to achieve conceptual coherence, and demonstrated care for their classroom and regional communities. In learning diverse sciences, they put into practice methods of inquiry matched to the nature of the challenge. Timely and local project work required inventive assemblies of disciplines. Children, adolescents, and adults learned to value the experience of learning. Their teachers anchored instruction in time-honored principles of "direct experience" and "present interest" (Herbart's and Rousseau's words, respectively).

In keeping with aims advocated by John Dewey and David Hawkins, social goals disciplined curricula from waveform energy and communication to horse family ancestry and climate history. In several instances, the experience of learning demonstrated the usefulness of understanding in the context of promoting the public good—weather forecasting in the local farming community, for example.

The Stories

Film art introduced paleontology to secondary students (the John Day Fossil Bed's *Impressions from the Past*) and geology to graduate ones (*Koyaanasquati*). Visual arts have clearly advanced these sciences, a clear-case illustration of how the characteristics of a discipline have adapted to the nature of the challenge.

The horse fossil venture took no haste to arrive at evolutionary conclusions. Instead, exercises drew the class into evaluating the reasonableness of evolutionary inferences given the limitations of the data. The lessons also extended the significance of horse fossil evolution to insights about changing climate and the history of these changes in the region where the students lived.

The idea of "deep respect for the present moment" embeds geologic abstraction in a context of value and stresses the role of human responsibility for the future—in place of the conventional portrayal of human history as a vanishingly small segment of a geologic time line. "Wicked problems" of climate change and species extinction apply multiple forms of expertise in social contexts. Concrete examples of place substituting for time and solutions cohering across scales in time and space underscore the integrity and power of thinking within a particular domain without any need to cast the thinking in generic terms.

Their Meanings

These messages concatenate, suggesting the contours of a new paradigm that first disaggregates, then reassembles disciplines. Subjects establish trust by making plain the criteria they follow to warrant knowledge. The advice is straightforward: know when to trust the source and know when it's fake. Never warrant the conclusions with a universal method—there'll be paradigms for judging once the data's done.

Subjects earn a place at the curriculum table by demonstrating their worth in solving problems, especially ones of social significance. As a result, the science taught from place to place and time to time varies and this variation complicates the goal of comparing achievement in the sciences from one nation to another—or, for that matter, one place to another. How children in United States compare to those in Finland on standardized scientific knowledge matters less than whether citizens know when to trust and when to doubt science that matters in their lives. Present circumstances call for a paradigm that connects school learning to the interpretation of local landscapes from multiple perspectives—geological, ecological, economical, historical, and cultural.

VALUES BEYOND STANDARDS

Valuing the products of inquiry that are consequential to society predicates appreciation of scientific plurality. Unified, generic characterizations that reduce the sciences to common methods mislead. Neither purporting to teach "the" scientific method nor aligning particular studies with generalized standards will suffice.

Divorced from social value, devoid of purposeful context, and deficient in coherence, science teaching stagnates. Adapted to local circumstances, embedded in social and political empowerment, and made responsive to cultural identity, science teaching may rejuvenate curricula beyond standards.

Liberation

Not only environmental hazards but also natural disasters have a social justice component. In the case of earthquake and tsunamis, helplessness engulfs many confronted with the possibility—and eventually, inevitability—of catastrophe. By acknowledging personal, social, and political actions that might ameliorate harm and hasten equitable recovery in the wake of a seismic catastrophe, citizens of coastal communities in the Pacific Northwest have taken small steps in the direction of liberation from helplessness.

Brazilian Paolo Freire's theory of pedagogy emphasizes becoming aware of the realities shaping life and developing the capacity to transform those realities. For Freire, education is a means of liberation from oppression perpetrated by elites and the forced construction of social reality serving their interests. The principal aim of education is to overcome powerlessness, achieved in two distinct stages. The first stage "unveils the world of oppression" to those who suffer most on its account and engages them in finding ways to reconstruct and hence transform realities that restrict their human potential and empower its liberation.

Having achieved awareness of oppression, its unwarranted rationalizations, and a shared commitment to change, Freire's pedagogy moves on to a second stage. "Pedagogy ceases to belong to the oppressed and becomes a pedagogy of all people in the process of permanent liberation."[9] In the second stage, dialogue unfolds across the community about how injustice limits human potential. Until social change unleashes this potential, science and technology have little prospect of resolving the wicked problems of collapsing biodiversity, climate change, food security, the degradation of marine environments, public health, disinformation campaigns . . . and on and on.

Capacity

Times have changed; Pleistocene dangers no longer confront society, and no one needs to learn how to club saber-tooth cats.[10] Equity, resiliency, sustainability, and capacity matter to society. Citizens deserve science education that teaches how to navigate these concerns with justice.

A preeminent social value of the twenty-first century is to build community capacity utilizing scientific knowledge—resiliency in the shadow of natural hazards, preservation of biodiversity in the wake of habitat loss, and economic vitality in the response to climate change. Sciences subordinated to practicing sustainable resource utilization and achieving equitable living standards deserve prominence in the curricula of public schools. For science education to thrive, the progress of its paradigm requires curricula that demonstrate their social value.

Quest

The stories shared in *Beyond Science Standards* from Oregon, Colorado, New York, China, Denmark, and Chile inspire hope. Teachers everywhere will recognize themselves and their aspirations for their students in these examples.

The rhetoric of school science conforms to a conception of inquiry that since 1630 has quested after certainty and universality. The expiration date for that quest has passed. The world of paradox, dilemma, conditionality, and uncertainty—the world of knowledge useful within limitations and responsive to valued aspirations—encapsulates the predicament of modernity. Truth-in-context has more to offer than the holy grail of absolutism. Science conceived in the singular and promoted as universal has warped the fabric of educational discourse. Science educators, enamored by visions of teaching science as lists of processes, methods, practices, or universal habits of mind, have difficulty seeing the light beyond the event horizon created by intense devotion to science standards.

The question, "What is the social purpose of science education?" has long endured. Is the answer to promote civic virtue, personal agency, or equal opportunity? To recognize, appreciate, and defer to scientific expertise? The purpose of education in the United States is to prepare citizens for democratic living. Each generation of Americans has faced this challenge in their own way and in response to changing historical circumstances of immigration, war, climate change, technological innovation, inequality, conspicuous consumption, discrimination, and economic crisis.

As voiced epoch after epoch, the value of scientific truth is to improve the human condition. Descartes certainly thought so, but historical circumstances, and the evolution of the sciences, have superseded his agenda of grounding claims about nature in universal certainty. That paradigm has run its course. Post-Cartesians satisfy themselves with finding compelling bases for agreement, context by context, while admitting to uncertainty.

The fields of school science need to be radically disaggregated, their fruits harvested according to their worth in solving societal problems, the curriculum varying from place to place, time to time, and purpose to purpose. That's the message the stories of Wavy Elephants, Binary Banjos, Balloon Eggs, Hell's Pig, Skull Sockets, Menacing Mudflats, Rock Poetry, Harvesting Oysters, and several more have to tell about progress in the quest for a new paradigm.

NOTES

1. Stephen Toulmin, *Cosmopolis: The Hidden Agenda of Modernity* (New York: The Free Press/Macmillan, 1990), 194.

2. Mario Livio, *Brilliant Blunders from Darwin to Einstein—Colossal Mistakes by Great Scientists That Changed our Understanding of Life and the Universe* (New York: Simon & Shuster, 2013), 263.

3. Simon Schama, *Landscape and Memory* (New York: Alfred A. Knopf, 1995), 12, 15.

4. John L. Rudolph, "The War on Method," in *How We Teach Science* (Cambridge, MA: Harvard University Press, 2019), 118–137. In this chapter Rudolph recounts how the "Harvard case studies" changed the conceptualization of science teaching and describes the role played by Thomas Kuhn, then a graduate student.

5. Charles R. Ault Jr., "Intelligently Wrong: The Meaning of Children's Misconceptions," *Science and Children* 21, no. 8 (May 1984): 22–24.

6. Jill Lepore J., *These Truths: A History of the United States* (New York: W. W. Norton, 2018), 390.

7. Ibid., 414–419.

8. Naomi Oreskes, *Why Trust Science?* (Princeton, NJ: Princeton University Press, 2019), 158.

9. Freire, *Pedagogy of the Oppressed*, 54. Paulo Freire, *Pedagogy of the Oppressed: 50th Anniversary Edition*, trans. Myra Bergman Ramos (1970; reis., Bloomsbury Academic, 2017), 54.

10. J. Abner Peddiwell, *The Saber-Tooth Curriculum* (New York: McGraw-Hill, 1939). Harold. R. W. Benjamin assembled his "lectures in the history of Paleolithic Education" into a venerable parody of educational practices that persist even though they no longer serve a useful purpose ("J. Abner Peddiwell" was his pseudonym).

Index

Page locators in italics refer to figures.

accommodation, 15–16
achievement gap, 288
Adams, Douglas, 3
aesthetic expression, xii, 5–6, 30, 149, 152, 192, 195–200, 220; and Horse Fossil Study Kit, 199–200; and iconography, 197–98, 201n6; as method, 159–60, 197; and understanding, 30, 160, 167–68, 188. *See also* art; illustration
African drumming and tonal languages, 70–71
Agassiz, Louis, 181–85
Age of Coal, 116
The Age of Reptiles (Zallinger), 190
Air Quality Egg System (AQES), 248–55, *249*, *253*, 285, 286
algae, red (branching coralline, *Amphiroa Orbygnyana*), *170*, 170–71, 175nn16, 20
algorithms, 32–33, 265–66, 297–98
Allmendinger, Richard W., 21–22
Allmon, Warren, 200
Alvarez hypothesis, 232n8
Amateur Ham Radio Club of Cornell, 250
American Association for the Advancement of Science (AAAS), 44
analogies, 11, 13, 16–17, 21–23, 220, 281; Darwin's use of, 166–68, 171–72; genetics and climate modeling, 260; in geologic inquiry, 97, 163–68, 229; of ice to rock, 97, 207–9, 212; past interpreted through analogy to present, 156, 163–65, 167–68, 227, 229–31, 270–71; scientific, to works of art, 151
anastomosis, 172
Anchiornis huxleyi (four-winged theropod), 196–97
Andes mountains, 21, 168
Angola, Kalahari Desert, 229
Anning, Mary, 142
anomalies, 16, 21–22, 96, 102, 111, 228
Anterior Orbit Point (AOP), 118–19
Anthropocene, 270–71, 275n18
Anthropologie (Haeckel), 193n9
Apollo Moon landing (1969), 272
Arbor School of Arts and Sciences, 237–44, 285, 286; automata projects, 238–41; celebrate everything, 238–39; "Inventions and Discoveries" theme, 239–43; "Letters to Aristotle," 243–44;

project proposals, 241–43; sharing, 243–44
Archaeopteryx, 178–81, *179*
Archaeotherium caninus (entelodont), 139–40
Archaeotherium Mortoni (Hell Pig), 139–41, *141*
Archimedes, 240
Archimedes's principle, 94–95, 97–98n4
Aristotelian principles, 25
Armadillidiidae, 189
"arms races," evolutionary, 130, 131, 133
art, xii–xiii; artistic expression, 6, 197, 200, 220, 244; artistry, criteria of, 150–52, 192; camera lucida, 152, 154n13; combined with science education, 89–90, 116, 152–53, 213, 219, 221, 238–39, 241–42, 298; and empathy, 150–52, 159; landscape, 147–49, 188, 215n17, 272; as method, 159–60, 197; school galleries, 187–88; stick figures sketched at zoo, 149–50; vivid depiction, 150, 152. *See also* aesthetic expression; Beijing Normal University; illustration
Arthropoda, 189–90
Art Walk trail (Cape Disappointment State Park), 213
Asia-Pacific Economic Cooperation forum, 205
Astor, John Jacob, 123n6
atmosphere, 81, 247–54, 265; carbon dioxide concentrations, 250, 270–71; and modeling, 258–60; sense of, 255
atolls, 167–68, 173–74n8
atomic bomb, as start of Anthropocene, 271, 275n18
Atwater, Brian F., 37–38, 46
Audubon, John James, 151–52, 154n14
Australia, 267, 271
automata, 237–45, 285; ant marionettes, 237; Robin Hood and Little John, 6, 238, 242; simple machine basics, 240–41. *See also* Arbor School of Arts and Sciences
avian dinosaurs, 180–81, 269
Bachman, John, 152
Bailey, Liberty Hyde, 100–101, 257
Bakker, Robert T., *179*
balloons, weather. *See* Bootleg Balloon program (Cornell University)
banjos: binary, 5, 65, 71–73, *72*; "squawking-hen" design, 66
Baum. L. Frank, 257, 261
Bay of Baiae (Gulf of Naples), 163, 165
Beagle, HMS (ship), 165–68, 195
beauty, 6, 43, 139, 159, 180, 192, 195; as problematic for physics, 32; sharing, 195–96
beaver family (*Myagaulidae*), 138
Beijing Normal University, 94–97; Gongshi (scholar's stones, spirit rocks), 206, 207, 212–13; Teach Less/Learn More program, 6, 205–6, 225; Thinking Geologic Thoughts workshop, 206–7, 225–26
bend metaphor, 13
Benjamin, Harold. R. W., 303n10
Beringia Land Bridge, 113
binary banjo exploration, xvi, 5, 65, 71–73, *72*
binary code, 5, 71–73
Biological Sciences Curriculum Study (BSCS), 91
birds, 180–81; *Archaeopteryx,* 178–80, *179*; as avian dinosaurs, 180–81, 269; and evolution of beauty, 195–96; melanosomes, 196; modern, 181
Birds of America (Audubon), 151–52
Bishop, Ellen Morris, 143
Bjornerud, Marcia, 218
block diagrams, 226, 227
Bony Hyoid Fish, 52
bookbinding, 239
books, printed, 157
Bootleg Balloon program (Cornell University), 247–55, 256n9, 258, 285, 286
Boston Legal (television show), 189–90, 193n6
branching diagrams, 168–72, *169*, *170*, 198

Bredekamp, Horst, 170
browser-grazer evolution, 113, 118–21, 125, 127–30, 158, 198
Brusatte, Steve, 180
Bryan, William Jennings, 297
buoyancy, 87–98n4, 93–97, 209
Bybee, Rodger, 91

Cajete, Gregory, 278
calderas, 144, 147–48, 230
California ground squirrel (*Otospermophilus beecheyi*), 102
Cambrian Period, 166, 189
cams, 241
canyons, 147
capacity building, 301
Cape Disappointment State Park (Washington), 213
Capra, Fritjof, xiii
carbon 12 and 13, 134, 158, 229
carbonate organisms, 167
carbon dioxide, 3, 133, 249–53, *253*, 262, 271
Carboniferous Period (Coal Age), 166, 269
Carrington, John F., 70
Carroll, Lewis, 212
Carroll, Sean, 32
Carson, Rachel, xiii
Cascade range, 114, 200, 230
Cascadia subduction zone, 37–41, *38*, *39*, 45, 110, 230–31
case studies, 2, 24, 25, 41, 303n4; examples of purposeful, 31–32
case thinking, 211
categories, 30, 43, 229; invention of, 209–11
cause and effect, 32–33
"causes now in operation," 164–65, 172, 230
Cecil (theropod statue, *Coelophysis*), 179, 188
celebration, 6, 237–39, 243–44
Celilo (town), 280
Cell Biology (journal), 170
Cenozoic Era, 109–10, 113–15, 125
"centers of creation" imagery, 184
certainty, quest for, 12–15, 34, 36n21, 43, 220, 296, 301–2
Chickaree *(Tamiasciurus douglasii),* 101
Chicxulub crater structure (Yucatán Peninsula), 24, 232n10, 268
child development, 15–16, 102–4
children: nature study approach with, 100–102; questions of, 85–86; as scientists, 80–81, 88–89; trauma from disasters, 45–46. *See also* students
Child Study Movement, 102–4
Chile, 168
China: Liaoning Province, 158, 196; Tari Basin, 229. *See also* Beijing Normal University
choice, 58, 86, 217; and coevolution, 196–97
Cisne, John, 190
citizens, 247–48; development of citizenship in students, xiv, 5, 261, 287, 290, 302; and knowledge of science, 13, 44–46, 251, 262, 277; responses to scientific discoveries, 39, 40, 251; students as citizen scientists, 251, 255, 289–90. *See also* community
clear, border, and counter-case thinking, 209–11
climate change, xii–xiii, 17, 33, 248–50, 252, 254–55, 266–73, 287; Anthropocene epoch added, 270–71, 275n18; carbon ratio of grasses, 134; decline in biodiversity, 267, 269–70, 298, 301; DNA comparison, 260; and evolutionary change, xii, 110, 114–17, 125–26, 131, 139, 153, 158; and global structure of circulation, 254; and horse evolution, 110, 114–17, 125, 128, 139, 158, 200; local effects of, 31, 261; modeling, 31, 88, 260, 266; paleontology and understanding of, 200
climate science, 3–4, 261–62, 266, 286; hybridized approach of, 33. *See also* Bootleg Balloon program (Cornell University)

Coelophysis (theropod), 179, 188–89
coevolution, 195–97
coherence, 5–6, 44, 159, 206, 298, 300; and credibility, 151; Herbartian, 104; and knowledge, 220–22, 225–26, 231; lack of in matrices and databases, 30; Teach Less/Learn More program, 205–6. *See also* understanding
Cold War, 295
Colorado Plateau, 212, 229
Colorado River, 3, 147, 219, 229, 279
Columbia River, 37, 123n6, 280–81
Columbia River Intertribal Fish Commission, 280
communication, 17, 244; elephant, 51–55; illustration as, 156–57, 220, 226; integrity of, 24; nonverbal, 157; and wave patterns, 68, 71. *See also* wave basics
community, 29; capacity building, 301; and celebration, 6, 237, 243–44; farming, 248, 250–51, 285, 286, 289, 298; and fossil discoveries, 142–43; and Herbart's theory of interest, 103; interpretive role of, 156; and landscapes, 30, 44, 277–78, 280–81, 288, 300; partnering with, 251; and rationality, 24–25; and resiliency, 40–41, 44–46; school as, 6, 44–46, 237, 244, 285–86. *See also* citizens
comparative thinking, 229
Comstock, Anna Botsford, 5, 104, 150; *Handbook of Nature Study,* 100–102, 182–83, 258, 273; and Panic of 1893, 257–58
Conant, James B., 13, 18n4
Concepción, Chile, 168
conceptual change theory, 2, 15–16, 18n3, 147, 223n4
conceptual coherence, 5, 6, 221, 298
conceptual invention, 91
concrete experience, 25
consciousness, 217
conspiracy theories, 266, 297–98
content, 4, 12, 244, 294–96, 298–302
context, xii, 4, 5, 12–13, 16–17, 158–59; and coherence, 151, 230–31; and geological inquiry, 220–22; historical, 5, 11, 16, 24, 36n21, 43–44, 230, 262, 302; purposeful, 30–32, 45, 89, 91, 220, 261, 286, 300; social, 12, 16, 279–80, 285, 299; and stories, 226, 230; timely, local, and particular sciences, 2, 13, 29, 30, 220, 273, 281, 286, 294, 299–300; valuing, 4–5, 25, 222
continental drift, 14–15
continuity, 13, 88, 156, 223n4
copper plates, 156, 157, 177
Coppola, Francis Ford, 206–7
"coralline" (*Amphiroa Orbygnyana*), *170*, 170–71, 175nn16, 20
corals, 167–68, 173–74n8, 175nn16, 20
core ideas, 4, 30, 294
Cornell Laboratory of Ornithology, 51
Cornell University, 100; Bootleg Balloon program, 247–55, 256n9, 258, 285, 286; Emergent Climate Research Laboratory, 258–61; meteorology program, 247–51; Science Leaflets program, 103
cosmos, 11, 297
Covid-19 pandemic (SARS-CoV-2), xii, xiii, 256n9, 267, 272–73; and backlash against science, 296–97; trust issues, 297–98
creationists, 198
credibility, 3, 151, 181, 192, 220, 229, 296
Cretaceous-Paleogene boundary, 268, 269, 271
Crichton, Michael, 178
Crooked River Basin (Oregon), 144, 230
Crooks, Ramsey, 123n6
crosscutting concepts, 4, 30, 294
crowdsourced modeling, 260–61, 286
crustaceans, 189
crystal formation, 75–76, 207–8, 228–29
Cultivate the Scientist in Every Child exhibit, 90–91

curriculum, 5, 220; and coherence, 151; design, 192, 231; integrated, 237–43; regional focus, 44–46, 71, 199; Science Curriculum Improvement Study (SCIS), 82n92, 91, 103. *See also* Horse Fossil Study Kit; standardization

dairy farm data, 250–51
dark side of science, 17
The Dark Side of the Earth (Wood), 14
Darwin, Charles, 6, 81, 99, 165–73; analogies, use of, 166–68, 171–72; extinction theories of, 170–71; illustrations by, 166, 168–71, *169*, *170*; on natural selection, 131–32; opposition to, 183–84; on plankton, 195; satirization of, 191–92; South American fossil horse tooth (*Equus curvidens*), 113–14; *Works* (*Formation of Vegetable Mould through the Action of Worms, with Observations of Their Habits,* 99–100, 191–92; *Notebook B, The Transmutation of Species, 169,* 169–70; *On the Origin of Species,* 168–69, 171, 191; *Structure and Evolution of Coral Reefs,* 167)
Darwins Korallen: Frühe Evolutionsmodelle und die Tradition der Naturgeschichte (Bredekamp), 170
date clam (*Lithodomus*), 164
da Vinci, Leonardo, xii–xiii, 242
Day, John, 123n6
Dayville High School (Oregon), 116–17, 139, 149, 299
decoding, 71–73
deduction, 42
Dee, John, 160n2
deep time, 217–19, 222
DeGarmo, Charles, 103
Deinonychus antirrohopus, 178–79, *179*
Denmark, Stevns Klint site, 268, 270
density, 94–97, 98n6
Derra, Mario, 180
Descartes, René, 1, 11–12, 18n2, 24–25, 40, 296; certainty, quest for, 12, 43
developmental homologies, 191
Devonian period (Age of Fishes), 166, 188, 269
Dewey, John, 41–44, 102, 192, 285, 295, 298; five-step heuristic, 42–43; oversimplification of thought of, 19n15, 41–42; on psychologized subject, 226; on school as community, 237; view of disciplines as problem-solving resources, 295
"Did you know..." questions, 102
difficulty of situation, 42
Digital Atlas of Ancient Life, 159
Diné language, 278–79
dinosaurs: birds as avian dinosaurs, 180–81, 269; *Coelophysis* (theropod), 179, 188–89; Mitchell Plesiosaur, 142–44; tyrannosaurs, 178–79; *Tyrannosaurus rex,* 188, 190
dioptrics, 77, 83n4
disease and pandemics, 272–73, 279. *See also* Covid-19 pandemic (SARS-CoV-2)
disunity of sciences, 1–3, 43–44, 293
DNA, 171, 260
Dr. De Soto (Steig), 198–99
Dromomeryx ("giraffe-deer"), 140
Dry Canyon (Oregon), 148–49
dung beetles (*Pallichnus dakotensis*), 139
Dürer, Albrecht, 157

Early Man (Time-Life), 190–91
earthquakes, 44–45; Cascadia subduction zone, 37–39, *38*, *39*, 45; changes in sea level, 164; episodic tremor and slip events, 40; Japan, 38–40, 46; magnitude scales, 39, 47n6; Nepal, April 2015, 45–46; Pozzuoli (1538), 164; Talcahuano and Concepción, Chile, 168
Eden metaphor, 221–22
educational psychology, 102–3
"eggs," atmospheric measuring instruments, 248–55, *249*, *253*, 285, 286

Einstein, Albert, 32–33
Eisner, Elliot, 150–52, 192
Elementary Science Study (ESS), 55, 71, *72*, 87, 103
elephants: infrasound communication, 51–54, 57; playful mimicry of in earthquake exercise, 37, 46
"Elephant's Child" (Kipling), 52–53, 85
embryonic development, 170–72, 183–85, 191–92, 193n9
Emergent Climate Research Laboratory (Cornell University), 258–61
Émile or On Education (Rousseau), 77, 79
empathy, 150–52, 159
Engage, Explore, Explain, Elaborate, Evaluate (5E model), 91
engraving, 157
Enlightenment, 11
entelodonts: *Archaeotherium caninus*, 139–40; *Archaeotherium mortoni* (Hell Pig), 139–41, *141*, 144
environment regulations, xiii
Eocene Epoch, 113, 115; Paleocene-Eocene boundary, 270–71
Eohippus (*Hyracotherium*), 117, 119, 198
Equidae, 113
Equine Evolution (film), 117
equity: competitive, 287–88; reciprocal, 287–90, 298; through schooling, 29
Essential Questions, 25, 55, 57, 114, 134
eugenics, 297
eurypterids (sea scorpions), 189
Evans, Connie, 137, 138
Evans, Robert, 270
Evidence as to Man's Place in Nature (Huxley), 191
evolution: aesthetic, 195–96; "arms races," 130, 131, 133; browser-grazer, 113, 118–21, 125, 127–30, 158, 198; and climate change, xii, 110, 114–17, 125–26, 131, 139, 153, 158; coevolution, 195–97; gradualist paradigm, 23; penny exercise for plotting, *126*, 126–30, *129*; primordial form, 171; punctuated equilibrium, 23–24; transmutation of species, 169–72, *170*, *171*. *See also* horses *(Equus)*, evolution of; Horse Fossil Study Kit
"Evolution: From Dinosaur to Bird" (Natural History Museum of Denmark), 180
The Evolution of Beauty (Prum), 195
experience, direct, 103–4
experience as teacher, 42–43
experimental design, 148
explanation, patterns of, 14
expressive language, 75–77
extinction: and Covid-19 pandemic, 267; Cretaceous Period, 24; Darwin's theories, 170–71; decline in biodiversity, 267, 269–70, 298, 301; mass extinction events, 267–69; meteor impact's effect on, 24, 220, 228, 232n8; patterns of, 23–24; as wicked problem, 267–72, 299

farming communities, 248, 250–51, 285, 286, 289, 298
Faulkner, William, 219
feeling, 71, 188, 213, 220–21; art and science projects, 187–88; as bridge between volition and cognition, 103, 285; and community, 103, 237, 281, 285; of satisfaction, 197, 221; of significance, 103, 151, 160, 244
felids, 130–31
Feyerabend, Paul, 25
fields, spacetime and quantum, 32
films, 114–17, 149, 206–7, 299
Fireman, Robert, 230
FitzRoy, Robert, 166, 168
five-step heuristic, 42–43
Formation of Vegetable Mould through the Action of Worms, with Observations of Their Habits (Darwin), 99–100, 191–92
Fossil, Oregon, 141–44
fossil fuel burning, 255
Fossil Horses (MacFadden), *112*, *118*, 121, *122*

On Fossil Objects, Chiefly Stones and Gems, Their Shapes and Appearances (Gesner), 155–57
fossil records, 33, 114; abrupt changes in, 267–68; dating practices, 120–21; of horse, 109–23; and punctuated equilibrium, 23–24; transition fossils, 127–30, 158, 198. *See also* paleontology
fossils: imagery of, 155–61; sixteenth-century understanding of, 155–56; virtual, 158–59
Four Horsemen, from the Apocalypse (Dürer), 157
frazil ice, 208–9
free-choice learning, 86
Freire, Paulo, 300–301
Fremd, Ted, 115, 121, 137, 138, 141
friction, 240
Funafuti, island of, 173–74n8

Galileo Galilei, 18n2, 239, 241
gawker science, 45–46
generic approaches, xiii, 1, 16, 18n4, 30, 41, 299–300
genetic variation, 200
genome, 171–72
genotype, 260
geography, reshaping of, 110
geologic inquiry, 21–22, 157, 217, 220–21; causal solutions, 227–28; "causes now in operation," 164–65, 172, 230; comparative thinking, 229; illustration necessary to, 226; nature of, 227–31; singular statements, 23, 26n10, 220, 227; and story form, 230–31. *See also* plate tectonics; volcanoes
geologic time, 218–19
geophysics, 37
geosciences, art in, 147
Gesner, Conrad, 155–57, 160
"Ghost Forest of Copalis," 38
Ghost Ranch, New Mexico, 179, 188
"Giant Insect and Fish Museum," 187–88
glaciations, 110, 113, 167, 184, 268, 270
glaciers, 208
global positioning systems (GPS), 32–33, 40
Glover, Danny, 217
Gomphoterium (elephant relative), 140
Gongshi (scholar's stones, spirit rocks), 206, 207, 212–13
Gould, Stephen Jay, 23–24, 197–98
Grand Canyon, 217, 226, 229; Trail of Time, 219, 222
Grand Canyon (movie), 217, 219
grand challenges, 33–34
graphing, three-dimensional, 127, 128
graphs, *60*, 60–61
grasses, C3 and C4 plants, 133–34
grasshoppers, 181–83, *182*
Grayling, Anthony C., 160n2
grids, 187, 242
Groenig, Matt, 191
Gruenewald, David, 277
Gubbio, Italy, iridium anomaly, 24, 232n8

habits of mind, 1, 22, 41, 44, 81, 90, 100, 294, 302
Haeckel, Ernst, 191, 193n9
Haemulon (scale-finned grunt fish), 184
Hall, Edwin, 184
Hancock Tree, (Oregon), 114
Handbook of Nature Study (Comstock), 100–102, 182–83, 258, 273
Hanna, Wyoming, 270–72
Harvard University, 2, 13
haste, as harmful, 55, 75–80, 82, 101, 211, 299
Hawaii, volcanic flows, 208
Hawkins, David, 4, 5, 80, 86–87, 90–91, 298
Hawkins, Frances, 86, 90–91
Hell Pig *(Archaeotherium Mortoni),* 139–41, *141*, 144
Henslow, John Stevens, 165
Herbart, Johann Friedrich, 103–4
Herbart and the Herbartians (DeGarmo), 103

historical sciences, 24
The Hitchhiker's Guide to the Galaxy (Adams), 3
Holocene, 270
The Holy Earth (Bailey), 100
hominids, 190–91
homologous skeletons, 111–12
hope, 221–22, 297–98, 301
Hopwood, Nick, 193n9
Horner, Leonard, 172
Horse Fossil Study Kit, 111–21, *112*, 139–40, 148–49, 198, 299; aesthetic components missing from, 199–200; Body Length chart, 125–30, *126*, *129*, 132; Molar Length for North American Horse Fossils chart, 132–34, *133*; as virtual fossils, 158–59; "You Are What You Eat" unit, 132–34. *See also* John Day Fossil Beds (Oregon)
horses *(Equus)*, evolution of, 6, 109–23, 198; body length evolution, *126*, 126–30, *129*; and browser-grazer evolution, 113, 118–21, 125, 127–30, 158, 198; cannon bones, 111; and climate change, 110, 114–17, 125, 128, 139, 158, 200; *Hyracotherium (Eohippus)*, 117, 119, 198; *Kalobatippus*, *112*, 121, 122, 129–30; *Merychippus*, *112*, 122, 129–30; *Miohippus*, *112*, 122, 129–30; *Parahippus*, *112*, 122, 129–30, 140; Phylogeny of the Equidae chart, *112*, 121, 129; predation and evolution of, 130–31; proto-horses, 115; South American fossil tooth *(Equus curvidens)*, 113–14; tooth morphology, 109–11, 115–23, *118*, *122*, 125–27, *126*; transition fossils, 127–30, 158, 198. *See also* John Day Fossil Beds (Oregon); tooth morphology
horsetail plant *(Equisetum)*, 115–16
Hossenfelder, Sabine, 32
How We Think (Dewey), 42
Hubble telescope, 197
human-centered meaning, 211
Huxley, Thomas H., 22, 31, 34–35n4, 42, 44, 191
hypocone, 109, 122, *122*
hypotheses, 3, 24, 180, 227, 271; about extinction, 268–69; Alvarez, 232n8; horse evolution, 117, 128; integration across scales, 24; selection of appropriate features, 148, 153
Hyracotherium (Eohippus), 117, 119, 198

ice: analogy to rock, 207–8; cube observation, 93–97, 98n5; frazil, 208–9
ice ages, 110, 113, 167, 184, 268, 270
icicles, 208–9
iconography, 197–98, 201n6
ideas, worth of, 2, 5, 18n3, 220–21, 223n4, 286, 295, 299, 302
illustration, xii, 6, 21, 147–49; as communication, 156–57, 220, 226; copper plates, 156, 157; Darwin's, 166, 168–71, *169*, *170*; engraving, 157; geologic maps, 14, 166, 226–27, 231; grids, 187; lithography, 177–78; as means of sharing information, 156, 177–78; and origin of paleontology, 155–56; sketching at zoo, 149–50; steel engravings, 157, 163; tree diagrams, 168–71, *169*, *170*; virtual fossils, 158–59; woodcuts, 155–57. *See also* aesthetic expression; art
imagery, 151, 168, 197–200; guiding images, 192; of human evolution, 190–92; racial, 184; of "structure," 23; tree diagram, 168–69, 171
imagination, 1, 82, 91, 152, 222, 242, 270
Impressions from the Past (film), 116–17, 139, 149, 299
indigenous peoples, 273, 278–82, 288–89, 293, 298; Salmon Camp, 279–81; Tsé na'alkaah students, 278–79
"Indigenous Physical Geology" course, 278–79

inferences, 42, 126–27, 207, 210, 299
information, transferred by waves, 54–61
infrasound, 52–54, 57
injustices, 17, 198, 272, 287, 301
inquiry, 16, 29–31; fruitfulness of, 16. *See also* geologic inquiry
instruments: chimes, 67–68; hammered dulcimer, 66–67; oven rack, 67–68; reeds, 69–70; spoon-chimes, 67; strings, 66–67; tubes, 68–69; whistles, 66. *See also* wave basics
integrated study, 1, 34, 71, 151, 188, 200, 239, 242–43, 273, 280, 286
"intelligently wrong," as useful to progress, 296
intelligibility, 16
interest, 89, 102–3
interference pattern, landscape as, 226–27
invertebrates, 163–75
iridium, 228, 268, 271
iron ore, 166
isopods, 189

Jackman, Wilbur, 102
Japan: earthquakes, eighteenth-century, 38–39; Tōhoku earthquake (2011), 40, 46
Jingshi Hotel lobby (China), 213–14
John Day Country (Oregon), 114–16, 120–21, 123n6, 128, 140, 150, 200; caldera, 147–48; Cant Ranch, 139
John Day Fossil Beds (Oregon), *112*, 114–16, 120–23, 125–30, 199; "The Cove," 140–41; *Impressions from the Past* (film), 116–17, 149, 299; Turtle Cove, 137–39. *See also* Horse Fossil Study Kit
Jordan, Teresa E., 21–22
journal entries, 3, 55, 60, 69, 101, 181–83, *182*; Darwin's, 113–14
journey, concept of, 13, 61
Jurassic Park (movie), 178, 190
Jurassic Period, 180, 269
Just So Stories (Kipling), 52–53, 85

Kalahari Desert (Angola), 229
Kalobatipppus, 112, 121, 122, 129–30
Karplus, Robert, 91
Kelsey, Frances Oldham, xiii
Kimmerer, Robin Wall, 288–89
Kipling, Rudyard, 52–53, 85
Kitts, David, 23, 24
Kline, Kevin, 217
knowledge: citizens and science, 13, 44–46, 251, 262, 277; and coherence, 220–22, 225–26, 231; and community, 244, 251; indigenous, 278–82; and paradigm shifts, 23; power, relationship to, 16–17; as relationships, 231; student construction of, 25, 61, 82, 240–42; and trust, 13, 17. *See also* understanding
Kolbert, Elizabeth, 267, 269
Koyaanisqatsi (film), 206–7, 299
Kuerbis, Paul, 105n17
Kuhn, Thomas, 2, 12–14, 16, 18n3, 24, 44, 303n4
Kyle, William F., Jr., 17

laboratory methods, 102–3
landscape art, 147–49, 188, 215n17, 272
landscapes, 222; and communities, 30, 44, 277–78, 280–81, 288, 300; geologic, 147–49; mapping, 226–27; and obsessions, 294; transformation of, 21, 54, 110, 115–16, 159, 167–68, 172, 227–28
Lascaux Caves, 159
lava, 14, 143, 166; ice compared with, 207–9, 214n8
lavacicles, 208–9
"leading thoughts," 102
learning how to learn, 94
learning progressions, 30
lecture/discussion, 86
legibility, 287, 290n2
length, *vs.* time, 218–19
Lepore, Jill, 297
Lerner, Neal, 184–85
Liaoning Province (China), 158, 196

liberation, 300–301
limestone, 164–67, 173; Colorado River theory, 229; Denmark, 267–68; karst, 167; lithographic, 152, 177; marine, at summit of Mt. Everest, 214, 227; Solnhofen, 178, 180, 185n3
Lin, Maya, 213, 215n17
lithography, 177–78, 180
locusts *(Pycnophlebia robusta),* 180
logical reasoning, 15
Look to the Mountain: An Ecology of Indigenous Education (Cajete), 278
Lopez, Barry, 277
Louis XIV (Sun King), 11
Lyell, Charles, 6, 163–68, 172–73, 178, 230

Maderspacher, Florian, 170
Madsen, Møller, 270
magma, 22, 143, 208–9, 212
magnification, 77
Magritte, René, 225
Maine, Neal, 289–90
Malaguzzi, Loris, 90–91
Malm, Rie Hjørnegaard, 270
mammals: Mastodon, 113–14; Megatherium, 113, 114; oreodonts, 115, 137–39, 159; rhinoceroses, 115, 138. *See also* elephants; entelodonts; horses *(Equus),* evolution of
Man in the Wilderness (Sklarski), 148
mapping, 14, 147, 166, 226–27, 231; art and data, 239; of weather, 254, 265
market paradigm, 288
Mascall Formation, 122, 140–41
Massachusetts Institute of Technology, 184–85
"Mass Tourism at the Twelve Apostles" (Derra), 180
Mastodon, 113–14
mathematical abstraction, 12, 32
Matthews, Michael R., 41
May, Rollo, xii
McPhee, John, 214, 217
meaning, 298–300; human-centered, 211; and worth of ideas, 220–21
meaningful learning, 6, 205, 220–21
Mediterranean Sea, 164
Megatherium, 113, 114
melanosomes, 196–97
Merychippus, 112, 122, 129–30, 140
Mesozoic Era, 180
messing about, 5–6, 86–91; analogues to rocks, 97; categorizing, 211; "Discovery" and "Exploration" symposiums, 87; echoes of, 90–91; with pendulums, 59–60, 63n14, 87–88. *See also* playfulness
Messing About in Science: Phases OΔ▢ (Hawkins), 86–87
metacognitive strategies, 209
metaphors, 147–48, 192, 231, 265, 277; ecological, 11; Eden, 221–22; geological, 218–21; length as, 218–19; "plates," 212; tree, 172; for understanding, 223n4
Metasequoia (dawn Redwood), 142
meteor impact, 24, 220, 228, 232n8
meteorology, 247–55; Air Quality Egg System and Bootleg Balloon program, 247–55, *248*, *253*, 256n9, 258, 285–86; algorithms, use of, 265–66; carbon dioxide sensors at ground level, 251–52; invisible processes, 253–55; mapping of weather, 254, 265; modeling for weather forecasts, 258–60; sense of atmosphere, 255; visualization tools, 254–55
method, art as, 159–60, 197
Miocene Epoch, 54, 121–22, 129–30, 141
Miohippus, 112, 122, 129–30
Mitchell Plesiosaur, 142–44
mixtures and solutions, 75–79, 82–83n2
modeling, 3–4; for climate change and weather forecasts, 31, 88, 254–55, 258–60, 265–66; computational forecasting, 265–66; crowdsourced, 260–61, 286; gumdrop models, 40
modernity, 11, 18n2
Montessori, Maria, 273
moose, 209–11

Moran, Thomas, 147
Morgan, Armand, *179*
Morris the Moose (Wiseman), 209–11
Morse code, 71
motion, physics of, 15
Mozart, Wolfgang Amadeus, 195–96
Mr. Potato Head, 81
Mt. Everest, 214, 227
multiple pathways, 86
Museo Carlos Darwin (Punta Alta, Argentina), 114
Museum of the Earth (Ithaca, New York), 179, 188–90, 248
museums, school galleries as, 187–88
Musical Instrument Recipe Book (ESS), 55, 71, *72*
mussel beds, 168
Mylagaulus (horned rodent), 140

National Science Foundation (NSF), 33–34, 248; *Elementary Science Study* (ESS), 55, 71, *72*, 87, 103; Science Curriculum Improvement Study (SCIS), 82n2, 91, 103
National Weather Service forecasts, 261
Native Americans, 278–82, 293, 298
Natural History Museum of Denmark, 180
natural selection, 130–32, 173
"Nature Closely Observed" (Hawkins), 80
"nature of the case," 3, 24–25, 31, 33, 220, 294, 296
nature study approach, 100–104, 182; and rural education, 257–58
Nature Study for the Common Schools (Jackman), 102
Navajo Nation, 278–79
neglect, constructive, 147–49, 151, 152, 192
"Nemesis" (companion star to Sun), 268
Nepal, April 2015 earthquake, 45–46
New Mexico, 188–89
Newport, Oregon, 40
Newton, Isaac, 18n2
Newtonian laws, 15–16
New Yorker, 267
Next Generation Science Standards, xii, 7, 25; Nature of Science section, 31; Physical Science Standard #4, 51, 54–61, 62n7; three "dimensions" of, 4, 17, 30, 44, 294. *See also* scientific method, "the"; standardization
nimravids, 116, 130, 139
North American tectonic plate, 110, 230–31
North Coast Land Conservancy, 289
Notebook B, The Transmutation of Species (Darwin), *169*, 169–70
Novak, Joseph, 6, 94, 205
Nova series (Public Broadcasting System), 191

Oates, David, 222
observation, 3, 33, 151, 156, 180, 184, 253–54, 266, 268, 270; Dewey's approach, 41, 42; and Horse Fossil Study, 118–20; and nature study, 100–103; selection of appropriate features, 148, 153
O'Keeffe, Georgia, 188
Old Red Sandstone, 165–66
Oligocene and Miocene epochs, 54, 115, 121–22, 129–30, *141*
oppressed, pedagogy of, 300–301
optics, 77–79, 83n4
optimistic responses, 46
Ordovician Period, 190
Oregon: Cascadian subduction zone, 37–41, *38*, *39*, 45, 110, 230–31; John Day Country, 114–16, 120–21, 123n6, 128, 140, 147–48, 150, 200; Smith Rocks State Park, 230–31. *See also* earthquakes; horses (*Equus*), evolution of; John Day Fossil Beds (Oregon); volcanoes
Oregon Paleolands Center, 142–44
oreodonts, 115, 137–39, 159
Oreskes, Naomi, 13
Ostrom, John, 178–80, *179*
oyster harvesting, 281

Pacific Northwest, 37–39, *38*, *39*, 45, 200; Salmon Camp, 279–81. *See also* Oregon
Pacific tectonic plate, 230–31
Page, Barbara, 189
Painted Hills, 137–39
Paleocene-Eocene boundary, 270–72
Paleocene-Eocene Thermal Maximum (PETM), 271
Paleocene Period, 269
paleoclimatology, 88
Paleontological Research Institution (Ithaca, New York), 159, 188, 200, 248
paleontology: collections, 126; origin of, 155–56; scaling, 117–18; visual underpinnings of, xii–xiii, 6. *See also* fossil records; Horse Fossil Study Kit
Panic of 1893, 257–58, 273
paradigm shifts, 2, 12, 14, 18n3, 24
Paradise Wild (Oates), 222
paradox, 152–53
Paradoxides gracilis (trilobite), 189
Parahippus, 112, 122, 129–30, 140
parental involvement, 55, 62n8, 72
particularity, 150–53
past, 217–19; interpreted through analogy to present, 156, 163–65, 167–68, 227, 229–31, 270–71
Peabody Museum of Natural History (Yale University), 178, 179, 190
Pedersen, Joel, 229
Penco, Italy, 164
pendulums, 59–60, 63n14, 87–88, 92n10
Perissodactyla, 115
Permian-Triassic boundary, 269
Pestalozzi, Johann Heinrich, 81
phenotype, 260
photosynthesis, 16
phylogeny, 113
Physical Science Standard #4 ("Waves and Their Applications in Technologies for Information Transfer"), 51, 54–61, 62n7
physics, 32–33, 293–94
physics, instrumentation in, 157
Physics Education Technology, 57
phytoliths, 133
Piaget, Jean, 15–16, 102–4
pill bugs, 189
pillow lava, 209
Pine Creek Schoolhouse (Fossil, Oregon), 141, 144
pitch, 65–70
place-based education, 29–30, 43–44, 237, 277–82
A Place of Discovery (film), 114–16
plankton, 195
plate tectonics, 14, 21, 40; analogues to rocks, 97; boundaries of plates, 230–31; density, buoyancy, and pressure, 212; North American and Pacific plates, 110, 143, 230–31; "plates" and "tectonic," as terms, 212; story line of, 212–14. *See also* earthquakes; volcanoes; winter ice tectonics
plausibility, 16, 69, 214, 270
playfulness, 30, 32, 138; at college level, 93–95; and disciplined thinking, 95–96; as innate, 81–82; mimicry of elephants in earthquake exercise, 37, 46; optics observation, 78–79; and present interest, 79; rainbow exploration, 89; and trust, 76; in wave basics explorations, 55–61, 65–73. *See also* messing about
Pleistocene Epoch, 113, 114, 122, 134, 189, 301
plesiosaur, 142–44
Pliocene Epoch, 116
Pliopithecus, 191
plurality, 22–23, 31, 296, 298, 300; singular-plural divide, 43–44; tolerating, 23–25
polis, 11, 297
politics, 2, 4, 273, 297
polygenism, 184
Port Desire, Argentina, 168
Portland, Oregon, 39
"A Portrait of Place" (Willapa Hills band), 281

Port Royal, Jamaica, 164
Posner, George, 15–16
Powell, John Wesley, 147
power, relation of knowledge to, 16–17
Pozzuoli, Naples, 163–65, 168
predation, 130–31
present, past interpreted through analogy to, 156, 163–65, 167–68, 227, 229–31, 270–71
present moment, deep respect for, 218–19, 221, 299
pressure, formula for ($P = \rho gh$), 96, 98n6
Principles of Geology: Attempt to Explain the Former Causes of the Earth's Surface by Reference to Causes Now in Operation (Lyell), 163–66, 173, 177–78
Principles of Paleontology (Raup and Stanley), 268
printers, 3D, 159
probabilistic reasoning, 126–67, 127, 265–67, 273
proboscideans, 53
process skills approach, xiii, 1, 31
progress, myth of, 17, 198
Project 2061: Science for All Americans (AAAS), 44
propositions, general, 42
Proterozoic, 219
Prum, Richard O., 195
public policy, 17, 265, 297
Punch's Almanack, 191–92
punctuated equilibrium, 23–24
Punta Alta, Argentina, 113–14

Quail Island, 166
qualitative methods, xii, 61
Quammen, David, 171–72
quantum fields, 32
questions, 85–86; about horse fossils, 110; in nature study approach, 101–2; "telling question," 89–90; "why" questions, 90

radiometric dating practices, 120–21
rainbow and mirror experiment, 89
Ratfish Wranglers (musical paleontologists), 141
rationality, 5, 11–12, 43, 156, 296; and community, 24–25
reality, 16, 151, 220, 231, 270; and coherence, 225; physics of, 32–33; socially constructed, 300; virtual, 247–49, 255, 266
reason, as response to centuries of war, 11
recitation methods, 103
Reggio, Godfrey, 206–7
Reggio Emilia schools, 90
regional focus, 21, 29, 45–46, 71, 199–200, 225–26; soil moisture measurement, 248–50. *See also* Bootleg Balloon program (Cornell University); Cascadia subduction zone; earthquakes; Horse Fossil Study Kit
relativity theory, 15, 32–33
religious agenda, 297
Renaissance, 239–41
renewal, capacity for, 221–22
representation, 57, 59–60, 190; anchored by metaphors, 231; geologic, 219–20, 225–27
Requiem for a Nun (Faulkner), 219
research communities, 24–25
resiliency, 40–41, 44–46, 301; of elephants, 54; optimistic responses, 46
responsibility, 220–22, 251, 266–67, 270, 282, 299; and democratic society, xiv, 40, 287–90, 302; development of citizenship in students, xiv, 5, 261, 287, 290, 302; and equity, 287–90; for the future, 221–22; social, 40–41, 272, 287, 290
reticulating patterns, 172
retrodictions, 23
retroviruses, 171–72
rhinoceroses, 115, 138
Richardson, Michael, 193n9
Ripple, Richard E., 103
river deltas, 229
rock, analogy to ice, 207–8
Rockcastle, Verne, 103, 105n17, 106n19

rocks and minerals, 213–14; basalt, 208; crystal formation, 207–8, 228–29; Elves Chasm gneiss, 219; feldspar, 76; Kaibab limestone, 219; Old Red Sandstone, 165–66; Solnhofen limestone, 178, 180, 185n3. *See also* limestone
Rocky Mountains, formation of, 21–22
Romey, William D., 147, 207–8
Roseman, Jo Ellin, 44
rote learning, 41–42, 79–80, 205, 231
Rousseau, Jean-Jacques, 77, 79, 90
Royal Geologic Society of London, 177
Royal Society Coral Reef Expedition of 1896-1898, 173–74n8
Rudolph, John L., 18n4, 19n15, 42, 295, 303n4
Rudwick, Martin, 155–56
rural education, 100–101, 257–58

saber-toothed cats, 116, 118, 130–31, 139
Salk Institute, 104n4
Salmon Camp, 279–81
San Andreas fault, 110
Sapp, Jan, 171–72
scale, 117–18, 230; and causal solutions, 227–28
Schama, Simon, 294
school: authoritarian schooling, xii, 41; as community, 6, 44–46, 237, 285–86; equity through schooling, 29; galleries made by students, 187–88. *See also* Arbor School of Arts and Sciences; Horse Fossil Study Kit; science education; *specific learning experiences*
Schulz, Kathryn, 39
Schwab, Joseph J., 23
"Science as Subject-Matter and as Method" (Dewey), 41
Science Curriculum Improvement Study (SCIS), 82n2, 91, 103
science denialism, xiii, 4, 250, 266, 296–98
science education: advancing schooling, 4–6; arts combined with, 89–90, 116, 152–53, 213, 219, 221, 238–39, 241–42, 298; automata and inventions, 237–45; certainty as focus of, 1, 34, 43, 296, 301–2; conceptual change theory, 15–16; integrated study, 1, 34, 71, 151, 188, 200, 239, 242–43, 273, 280, 286; pivot in, 4, 295. *See also* Next Generation Science Standards; standardization
"scientific," as adjective, 2
scientific agendas, 12, 24–25, 33–34, 43, 225, 293–94, 302
scientific diversity, xiv, 29–30, 296–97; disunity of sciences, 1–3, 43–44, 293; geologic inquiry as example of, 227; "natures of the sciences," 22–23; as resource for solving societal problems, 5, 31–32, 42–44, 295–96, 302
scientific method, "the": codified into short list of steps, 41–42, 295; Deweyan thought about, 41–43; as impoverished vision, 16–17; as myth, 2, 4–5, 17; as phrase, 1; remodel of, 29–30; scientists' efforts to educate public about, 2, 13, 18n4, 19n15, 295. *See also* Next Generation Science Standards
scientific practices, 4, 29–30, 294
scientific revolutions, 2, 12–13
SCIS Learning Cycle, 91
Scott, James C., 290n2
Scudder, Samuel, 184–85
seafloor, 14–15, 166, 207, 212, 229
sea level, 33, 163–65, 167–68, 174n8, 214, 270
Searle, Mike, 227
Seaside High School (Seaside, Oregon), 289
seasons, rhythm of, 251–53
Sedgwick, Adam, 165–66
Semken, Steven, 278
Senefelder, Alois, 177, 180
Serapis, Temple of (Pozzuoli), 163–66, 168, 173

sexual selection, as aesthetic evolution, 195–96
Shakespeare, William, 160n2, 241
Shatner, William, 189–90, 193n6
Sierras Pampeanas, 22
significance, feeling of, 103, 151, 160, 244, 285
simple machines, 6, 238, 240–41, 285, 286
The Simpsons (television show), 191
singularity paradigm, 31
singular-plural divide, 43–44
singular statements, 23, 26n10, 220, 227
"Sinking Ice Lesson," 93–97
Sitka spruce trees, 38, *38*, *39*
Skamokawa, Washington, 281
skepticism, 12, 156, 195, 198, 296
Sklarski, Bonnie, 148
Slinky experiments, 56–57
Smith, Edward, 16
Smith Rocks State Park (Oregon), 144, 230–31
Smolin, Lee, 32
snailery, 101–2
social interaction, 103, 250–51
social justice, 297, 300–301
social value, xii, 5, 12, 289, 294, 300–301; precious lives, 45–46; public trust in science as, 261–62; resiliency, 40–41
soil moisture, measurement of, 248–52
"solid solution," 75–76
Solnhofen limestone, 178, 180, 185n3
South American mountain building, 21–22
Soviet Union space program, 272, 295
sow bugs, 189
spacetime, 32
Spanish Armada, 160n2
Spanish flu (1918), 272, 297
Sputnik satellite (1957), 272, 295, 297
standardization, xii; and competitive equity, 287; cross-national comparisons of student achievement, 17, 29, 299–300; deficiencies of, 30, 41–44; and haste, 76–84; limited lists of core ideas, 34; momentum of movement, 29–30. *See also* curriculum; Next Generation Science Standards
"Standing Wave Maker," 60, *60*
steel engravings, 157, 163, 177
Steig, William, 198–99
stereotypical science, 4, 17, 23, 31, 293–96
Stevns Klint (Denmark), 268, 270, 271
stick figures, drawing, 149–50
stick in cup of water observation, 77–79, 83n4
stories, 5–6, 298–300; context provided by, 226; Elephant's Child, 52–54, 85; and geologic inquiry, 230–31
stratigraphic cross sections, 226, 227
string theory, 32
Strong Museum of Play (Rochester, New York), 81–82
"structure," 23
Structure and Evolution of Coral Reefs (Darwin), 167
Structure of Scientific Revolutions (Kuhn), 12–13
students: as citizen scientists, 251, 255, 289–90; development of citizenship in, xiv, 5, 261, 287, 290, 302; knowledge, construction of, 25, 61, 82, 240–42; Tsé na'alkaah, 278–79. *See also* children; *specific learning experiences*
study movements, 100–104
style, 23–24
subsidence and uplift, 14, 164, 166–68, 174n8. *See also* Cascadia subduction zone
Sun, possible companion star to, 268
sustainability, 17, 25, 279, 282, 301
sweat bees, 139
syncytin genes, 171–72

Tahiti, 174n8
Talcahuano, Chile, 168
The Talking Drums of Africa (Carrington), 70

Teacher Institute for Evolutionary Science, 191
Teach Less/Learn More, 6, 205–6, 225
The Tempest (Shakespeare), 160n2
Tephrocyon ("bear-dog"), 140
Terminator Pig. *See* Hell Pig *(Archaeotherium Mortoni)*
Tertiary Era (Age of Mammals), 268
On the Origin of Species (Darwin), 168–69, 171, 191
thermal maximum event, 271
These Truths (Lepore), 297
Thier, Herbert D., 91
Thinking Geologic Thoughts workshop, 206–7, 225–26
Thirty Years' War, 11
Thomas Condon Paleontology Center (Picture Gorge, Oregon), 116, 148
threats, obsession with, 295
Through the Looking Glass (Carroll), 212
Tibetan Plateau, 97, 212
Tierra del Feugo, 166
time, 32–33; deep, 217–19, 222; geologic, 166, 173, 184, 199–200, 212, 217–22, 230, 268–70, 299; length *vs.,* 218–19; visual representations of, 226
Timefulness: How Thinking Like a Geologist Can Help Save The World (Bjornerud), 218
time lines, 120–21, 218–19; student, 239
timely, local, and particular sciences, 2, 13, 30, 220, 273, 281, 286, 294, 299–300; causal solutions in geologic inquiry, 227–28; and climate science, 29, 33
"Timing is Everything" (Teacher Institute for Evolutionary Science), 191
Tlingit people, 278
Tōhoku earthquake (2011, Japan), 40, 46
Tolkien, J.R.R., 12, 260
tooth morphology: categorization scheme, 138; and emerging grasslands, 131–34; of horses, 109–11, 115–23, *118*, *122*, 125–27, *126*; of oreodonts, 138–39; and tooth chemistry, 133–34, 158. *See also* horses *(Equus),* evolution of
topography (landscape), 226–27
tornadoes, 253–54
Toulmin, Stephen, 12, 17, 18nn2, 3, 22–25, 36n21, 44; evolutionary metaphor for understanding, 223n4; on "nature of the case," 220; sciences respond to present circumstances, 262
tourism, collapse of, 267
Toy Story, 81
Trail of Time (Grand Canyon), 219
Transactions (Royal Geologic Society of London), 177
transition fossils, 127–30, 158, 198
transmutation of species, 169–72, *170*, *171*
tree diagrams, 168–72, *169*, *170*, 198
tree of life image, 170–71, 173
"Trees: Dead or Alive" (Cornell cooperative extension pamphlet), 103
Triassic Period, 188, 269
trilobites, 188–90; *Paradoxides,* 189; *Triarthus eatoni,* 190
trust, 3–5, 17, 43–44; and backlash, 296–97; and community, 24, 46; cultivating, 261–62, 296–98; and playfulness, 76; and reciprocal equity, 288
Tsé na'alkaah students, 278–79
tsunamis, 300; Cascadian area, *38*, 38–39, *39*, 41, 46, 289; Kathmandu, 44–45
tuffs (volcanic deposits), 120–21, 138, 143, 230
Twelve Apostles rock formation (Australia), 180
tyrannosaurs, 178–79
Tyrannosaurus rex, 188, 190

uncertainty, importance of, 3, 31, 33–34, 36n21, 302
understanding, 1, 6, 25, 40, 43, 199–200, 223n4, 225; and aesthetic expression, 30–31, 160, 167–68, 188; of geologic time, 217–22; and haste,

76, 80; insecure, 76; paradigm-shifting, 4; perceptions of patterns, 61; as pleasurable, 103, 192, 197, 220–21; public, 297–98; and serious playfulness, 76, 80–82, 85–86, 88; sophisticated, 76, 80–81; ungulates, 110. *See also* horses
unsettling, 208; of wave basics, 54–55, 59, 61. *See also* coherence; knowledge
uniformitarianism, 165, 173
universal approaches: assumption of, 1–2; and case-by-case search for truths, 25; certainty, quest for, 12–15, 34, 36n21, 43, 220, 296, 301–2; of Dewey, 41; Huxley's view, 22, 31, 42; as "one method to rule them all," 2, 12, 260; paradigm shift as counter to, 13; *vs.* particularity, 150–53; physics and search for, 293–94; of Piagetian paradigm, 15; singular-plural divide, 43–44
University of Chicago, 102
University of Copenhagen, 180, 270
Upadhyay, Bhaskar, 45

Vale of Clwyd (Wales), 166
Valparaiso, Chile, 164
verification, 42
vertical evacuation platforms, 46
vibrations, 55–56; controlling and decoding, 71–73; pitch, 65–70; sensed by worms, 99
Vietnam Veterans Memorial (Washington, DC), 215n17
virtual earths, 3
virtual particles, 32
virtual reality goggles, 248, 249
viruses, 171–72
viscosity, 211
visualization, 155–56, 199, 226, 231; for meteorology, 254–55
vocabulary, 75–76
volcanoes, 14–15, 143–44, 269, 271; Cascade range, 230; Chile, 21–22; and fossil record, 109–10, 115, 120–21, 143–44, 199; Hawaii, 208; John Day Country, 139, 141, 143–44, 147–49, 159, 199–200; Smith Rocks State Park, 230–31; tuffs (volcanic deposits), 120–21, 138, 143, 230
volume, 94–96, 98n6
Voss, Julian, 170

war, influence on thinking, 11, 12
wave basics, 54–61, 68, 71; dots in pan of water exploration, 58–59; energy, 59; frequency, 57; graphs, *60*, 60–61; interference, 58; nodes/antinodes, 58; outdoor explorations, 59; overtones, 67; pitching sound, 65–70; pulses, 57–58; resonance, 59; "saltoscilloscope" exploration, 59–60; speed/tension, 58; squawking-hen system, 65; "Standing Wave Maker," 60, *60*; vibrations, 55–56; "wave on a string" simulation, 57, 68n12. *See also* communication; instruments
weather. *See* meteorology
Weather Research Forecast software, 260–61
Weather Truck, 247–48, 286
Wegener, Alfred, 14
West, Elliot, 278
whippos (whippomorpha), 140
Whitehead, Alfred North, 6, 225
Wicked Device, LLC, 248, 251
"wicked problems," 251, 273n1, 299; climate change as, 265–67; Covid-19 pandemic as, 272–73; extinction as, 267–72
Willapa Hills band, 281
Williams, Larry, 142
Wilson, Wilford M., 258
Wilson, Woodrow, 297
winter ice tectonics, 209
The Wizard of Oz (Baum), 257, 261
Wonderful Life (Gould), 197–98
Wood, Robert Muir, 14
woodcuts, 155–57, 177
World War I, 297

World War II, 2, 14, 271, 273, 281, 295
worms, 99–100, 104n4, 191–92
worth of ideas, 2, 5, 18n3, 220–21, 223n4, 286, 295, 299, 302
Wu Jinshan, 94, 205–6, 225

Xi Jinping, 205

Yale University, 179
Yellowstone Caldera, 144, 231
Yellowstone National Park, 144
"Young Citizen Development Centers," 290
Young's Bay (Oregon), 37

Zallinger, Rudolph, 190–91
zebrafish, mutated, 185
Z hypothesis, 117–18, *118*
zoos, sketching at, 149–50

About the Author

Charles R. Ault Jr. ("Kip"), professor emeritus, Lewis & Clark Graduate School of Education and Counseling, coordinated the school's Science Master of Arts in Teaching program for more than two decades. Kip began his career teaching primary grades and middle school science in Connecticut and Colorado. Author of *Do Elephants Have Knees?* and *Challenging Science Standards*, his writing reflects an interest in paleontology from graduate study at Cornell University that led to valuing subject diversity in the reform of school science.

www.ingramcontent.com/pod-product-compliance
Ingram Content Group UK Ltd.
Pitfield, Milton Keynes, MK11 3LW, UK
UKHW021905150926
475560UK00003B/199